Routledge Revivals

Atmospheric Turbulence

Originally published in 1955 *Atmospheric Turbulence* examines dynamic meteorology and the fundamental part it plays in the overall science of meteorology. The book examines the theory of atmospheric turbulence as a more mathematically developed area than largescale motions of the atmosphere and examines its significance in economic, military and industrial spheres. The book focuses on the effect and importance of atmospheric turbulence, not only to meteorologists, but the designers of large aircraft. The book addresses the effects of turbulence and the properties of the atmosphere that can be found closer to the ground. This book will be of interest to atmospheric physicists and meteorologists.

Atmospheric Turbulence

by O.G. Sutton

LONDON AND NEW YORK

First published in 1955
by Methuen

This edition first published in 2019 by Routledge
2 Park Square, Milton Park, Abingdon, Oxon, OX14 4RN
and by Routledge
605 Third Avenue, New York, NY 10017

First issued in paperback 2021

Routledge is an imprint of the Taylor & Francis Group, an informa business

Publisher's Note
The publisher has gone to great lengths to ensure the quality of this reprint but points out that some imperfections in the original copies may be apparent.

Disclaimer
The publisher has made every effort to trace copyright holders and welcomes correspondence from those they have been unable to contact.

A Library of Congress record exists under LCCN: 55003055

Publisher's Note
The publisher has gone to great lengths to ensure the quality of this reprint but points out that some imperfections in the original copies may be apparent.

ISBN 13: 978-0-367-34018-6 (pbk)
ISBN 13: 978-0-367-34017-9 (hbk)

METHUEN'S
MONOGRAPHS ON
PHYSICAL SUBJECTS

General Editor: B. L. WORSNOP, B.SC., PH.D.

ATMOSPHERIC TURBULENCE

ATMOSPHERIC TURBULENCE

by

O. G. SUTTON

C.B.E., D.Sc., F.R.S.

DIRECTOR OF THE METEOROLOGICAL OFFICE

WITH 4 DIAGRAMS

LONDON: METHUEN & CO. LTD.

NEW YORK: JOHN WILEY & SONS, INC.

First published September 15th 1949
Second Edition, 1955

2.1
CATALOGUE NO. 4044/U

PRINTED IN GREAT BRITAIN

PREFACE

IN this Monograph I have attempted an account of an aspect of dynamical meteorology which is now recognized as a study of major importance, not only because of its intrinsic interest and the fundamental part which it plays in the science of meteorology as a whole, but also because of its significance in economic, military and industrial spheres. The theory of atmospheric turbulence is more mathematically developed than those parts of dynamical meteorology which treat of large-scale motions of the atmosphere and I have considered the subject throughout as a branch of mathematical physics. This is a partial and perhaps a biased view of the subject, and it is with real regret that for reasons of space I have had to resist the temptation to include accounts of the special instruments which have been devised, with ingenuity and skill, for the highly accurate observations which the subject demands. I have also refrained from discussing turbulence in the upper air, but for quite another reason. This aspect of the subject is of the greatest importance both to meteorologists and to the designers of large aircraft, but the information available as yet is meagre and hardly in a state suitable for treatment in book form. The Monograph therefore deals exclusively with the properties of the atmosphere near the ground, undeniably the most important region for turbulence. It is obvious that a book of this size can make no claim to be exhaustive, and my aim throughout has been to indicate to the reader those parts of the subject which seem to me to be significant for future development rather than to provide a long list of researches.

I am glad to acknowledge my indebtedness to Professor Brunt's *Physical and Dynamical Meteorology*, a fact which will be obvious to the reader by the numerous references scattered throughout the book. I owe a special debt to Mr. K. L. Calder, Head of the Meteorology Section at the Chemical Defence Experimental Station, Porton, who has been kind enough to read the book in manuscript and whose suggestions I have almost invariably adopted. To my secretary, Mrs. F. D. Crockford, I owe my thanks for the care she has shown in the preparation of the typescript.

MILITARY COLLEGE OF SCIENCE
SHRIVENHAM, BERKS.

March, 1948

PREFACE TO THE SECOND EDITION

THE six years which have elapsed since this book was written have seen one major development, the introduction of the Kolmogoroff statistical theory. I have added a brief account of the fundamental ideas of this theory, and short additional paragraphs on natural convection and atmospheric pollution to bring the book up to date.

I have to thank correspondents for pointing out some misprints, all of which have been corrected in this edition.

O. G. S.

June, 1954.

CONTENTS

CHAPTER I

TURBULENCE IN GENERAL

INTRODUCTION

A SENSITIVE anemometer placed anywhere within a few hundred feet of the surface of the earth shows that the motion of the air consists for the most part of a succession of gusts and lulls accompanied by rapid and irregular alternations in direction. This feature, immediately obvious to anyone watching smoke from a chimney or ripples passing over a field of corn in summer, is more than a meteorological curiosity. The unsteadiness of the wind has much to do with the shape of life as we know it, for it is this property which largely controls such apparently dissimilar phenomena as the warming of the atmosphere near the surface of the earth, the evaporation of water from land and sea, the scattering of pollen and the lighter seeds and (most important in an industrial age) the removal of pollution from the air above great cities and crowded centres of industry.

The systematic study of *atmospheric turbulence*, i.e. of the eddying of the wind as a diffusing process, is of recent origin and all the major developments have occurred within the last thirty or forty years. To a large extent the subject has advanced with the study of aerodynamics and any systematic account must therefore start with a background of fluid motion theory.

1. VISCOSITY

All fluids, whether liquids or gases, exhibit some resistance to deformation, or *viscosity*. The analysis of viscosity may proceed in two ways: (*a*) by considering the mechanical properties of a fluid continuum without reference to its internal constitution, and (*b*) by ascribing a molecular structure to the fluid and making use of the statistical properties of an assembly of molecules. An examination of both approaches is essential as a preliminary to the study of the problem of turbulence.

Viscosity as a Bulk Property. Consider a fluid initially at rest between two parallel planes, one of which is set in motion relative to the other. The fluid adheres to both planes and a resistance is experienced as the motion of the plane spreads to the fluid itself. If z is distance measured from, say, the fixed plane and $u=u(z)$ is the velocity of the fluid at any intermediate level we have $u=0$ on $z=0$, $u=U$ on $z=Z$, where Z is the distance between the planes and U is the velocity of the moving plane relative to the fixed plane. If for the present we restrict ourselves to very low speeds or small values of Z or highly viscous fluids we can explain what happens by supposing that between the boundaries there is set up a variable force, of the nature of a stress and always acting parallel to the direction of the motion. Experiment shows that the magnitude of this force is determined by the nature of the fluid and the rate of change of velocity with distance from a boundary. We are thus led to define a *shearing stress per unit area* τ as

$$\tau=\mu \, du/dz \qquad . \quad . \quad . \quad 1.1$$

where μ is a constant representing an intrinsic property of the fluid and called the *dynamic viscosity* $(ML^{-1}T^{-1})$. The quantity which normally enters into fluid motion calculations is, however, the *kinematic viscosity* ν, defined as the quotient of the dynamic viscosity and the density. For air near the ground $\nu=0{\cdot}15$ cm.^{2}sec.$^{-1}$ approximately.

The arrangement considered above, known as Couette flow, constitutes the simplest case of viscous motion but merits closer attention because it illustrates certain recurrent features of our problems. In the first place it should be observed that the principal effect of the viscosity is to transfer some part of the motion of the boundary to the fluid itself—in other words, viscosity causes a *diffusion of momentum* throughout the fluid. Secondly, the resulting motion is not uniform but continuously variable from one boundary to another, giving rise to a so-called *velocity profile* $u=u(z)$

which in the simple case we have considered is linear but which is in general curved. Finally, in the special circumstances considered (low speeds, shallow layers or high viscosities) the shearing stress is completely expressed by the product of the *velocity gradient* du/dz and a coefficient μ which is independent of u and z.

Viscosity as a Molecular Property. In considering viscosity as a consequence of the molecular structure of the fluid we confine our attention to the analysis of the viscous property for gases (due originally to Maxwell (1860)) in its simplest form.

The molecules of the gas are regarded as extremely small discrete masses moving at random, but with average velocity components equal along any set of axes. As before, the gas as a whole is assumed to be moving slowly between parallel plane boundaries, the bulk motion being too small to affect the distribution of molecular velocity. Let l be the *mean free path*, i.e. the average distance over which a molecule travels between collisions; l then measures roughly the thickness of a layer in which viscous action can take place and also the distance over which the molecule conserves its momentum.

Suppose that in the steady state a bulk velocity profile $u=u(z)$ exists in the gas and consider a layer between the levels z and $z+l$. Molecules situated on $z+l$ have not only their own high velocity c but are drifting with the main stream velocity $u(z+l)$. A typical molecule of mass m at this level will thus have an ordered momentum $mu(z+l)$ in the direction of the main current in addition to its random heat motion. If, as a result of the molecular agitation, a molecule crosses the plane $z+l$ at right angles, it will retain its momentum unaltered until it reaches the plane z where the typical molecule has an additional momentum $mu(z)$ in the direction of the main stream. The net result is the transfer from one level to another of a quantity of momentum equal to the difference $mu(z+l)-mu(z)=lm\, du/dz$ approximately. On the average, since for the molecules all directions of motion are equally possible, the number moving normal to unit area of the layer in unit time will be $\frac{1}{6}Nc$ in each direction, where N is number of molecules per unit volume, and the total rate of momentum transfer across the layer is therefore $\frac{1}{3}Nmcl\, du/dz$ per unit area.

Since conditions are steady, momentum must be transferred as fast as it is received so that the process is continuous right down to the stationary boundary on which a force $+\tau$ is exerted per unit area while an equal and opposite force $-\tau$ is exerted on the moving plane. Hence at all levels we have a shearing stress given by

$$\tau=\tfrac{1}{3}Nmcl\, du/dz,$$

or, comparing with 1.1 above, since $Nm=\rho$, the density of the gas,

$$\mu=\tfrac{1}{3}\rho cl, \qquad . \quad . \quad . \quad 1.11$$

the usual kinetic theory expression for the coefficient of viscosity. The argument is, of course, capable of considerable refinement which need not concern us here.

The essential feature of the kinetic theory analysis of viscosity is that in a non-uniform velocity field the effect of the incessant and random motion of the molecules is to transfer momentum from regions of high bulk velocity to regions of low bulk velocity. The same process may be imagined to apply to heat and mass, giving rise to the phenomena of *conduction* and *diffusion*. In all these cases the effect of molecular agitation is to cause *mixing*.

If molecular forces alone came into play mixing would necessarily be a very slow and localized process. It is a common feature of daily life that efficient mixing, whether it be of the molecules of sugar and water in a cup of tea or of the air in a room fitted with a fan is easily accomplished, and it is also evident that rapid and large-scale natural mixing must go on incessantly in the lower atmosphere. The means whereby such mixing is effected are obviously macroscopic, consisting essentially of the movements of large masses of fluid over considerable distances, as opposed to the microscopic and spatially limited action of the molecular agitation and the question immediately arises whether such macroscopic diffusion is sufficiently regular to exhibit laws, necessarily statistical, analogous to those established for molecular diffusion. This is the genesis of the turbulence problem.

2. TURBULENCE ON THE LABORATORY SCALE

The Couette flow considered above is characterized by the property that the layers of fluid glide over each other in a series of parallel planes, the only mingling of the fluid in adjacent planes being by molecular agitation. In a fluid of small viscosity, such as air, motion of this type can only be maintained permanently in very special circumstances which will be explained later.

Reynolds' Experiments. In 1883 Osborne Reynolds published the account of his now-classical experiments on flow in long straight pipes. In those experiments the motion of the liquid was made visible by the introduction of dye through a small auxiliary tube and Reynolds was able to demonstrate the existence of two widely different modes of motion, namely, *laminar* (or *streamline*), when the thread of dye moved downstream with little or no visible change, and *turbulent* (or *eddying*), evinced by the rapid break-up of the filament of dye so that the tube quickly became filled with dilute colour. Reynolds also concluded, by dimensional analysis, that the transition from laminar to turbulent flow depends only on the value of the dimensionless combination ur/ν, where u is the average velocity of the fluid, r is the radius of the pipe and ν is the kinematic viscosity of the fluid. This quantity is now called the *Reynolds number* (*Re*) and is the basic parameter of modern aerodynamical theory.

Reynolds' experiment has been the subject of many subsequent investigations, both practical and theoretical. His principal result is often stated in the form that for pipe flow a value of *Re* about 1,000 is critical in the sense that above this the motion is turbulent, and below it, laminar. This statement, however, requires qualification. There appears to be no definite upper limit for *Re* above which fluid motion is invariably turbulent—by taking pains to obtain a smooth entry and by insulating the apparatus from all external disturbances Ekman (1910) observed laminar flow as high as $Re = 12{,}000$. On the other hand, it seems to be well established that no matter how great the initial disturbances, turbulence cannot persist in a long pipe if $Re < 1{,}000$.

The above considerations relate to conditions in which the density is uniform. In the case of a fluid heated or cooled from below there is a gradient of density in the vertical, and the onset of turbulence depends, in addition, on the influence of gravitational forces. As we shall see later, this situation is of frequent occurrence in the lower atmosphere and is responsible for many of the difficulties met with in the meteorological case, but consideration of this intricate problem is deferred until a later chapter.

On the theoretical side attempts have been made by a number of distinguished mathematicians, including Reynolds, Orr, Rayleigh, Heisenberg, Tietjens, Southwell and Chitty, Tollmien and Schlichting, to examine theoretically the transition from one type of flow to the other. The method usually adopted has been to investigate the conditions in which a disturbance of given form, imposed on a basic laminar flow of known characteristics, increases or decreases with time so that the problem is regarded as essentially one of stability. These investigations have invariably proved to be long and difficult and to date have not succeeded in establishing a completely rational basis for a theory of turbulence. In a book of this size it is impossible to give an adequate discussion of the matter and the reader must be referred to advanced treatises on fluid motion for further information.*

3. THE NATURE AND DEFINITION OF TURBULENCE

Flow in long straight pipes resembles in many ways the motion of the air near the surface of the earth and has the advantage that its characteristic features have been measured with a precision only possible in the laboratory. If a sensitive instrument of quick response, such as a hot-wire anemometer, is used to measure velocity in the turbulent state the record reveals a very irregular motion with no evidence of periodicity. In this respect the contrast between laminar and turbulent flow is not unlike that between a pure musical note and a 'noise' in acoustics.

* A review of this subject is also given in C.S.I.R., Australia, Division of Aeronautics Report A. 35 (1945), by A. F. Pillow.

Mean Properties. At this point it is necessary to introduce a concept of major importance and which will be defined with greater precision later. It is obviously extremely difficult, if not impossible, to formulate anything except statistical laws for the irregular instantaneous velocity, and attention has been primarily concentrated on *mean values*, usually averages over a period of time. The use of such means is not restricted to velocity but covers other properties such as temperature, or concentration of suspended matter, which themselves exhibit fluctuations in a turbulent fluid. Usually, the mean value is conveniently and automatically obtained by the use of an instrument of slow response (e.g. by a pitot tube connected to a manometer, or by a fairly sluggish thermometer), and although such instruments do not give the true mean, the error can usually be made negligible by a suitable choice of the instrumental constants. We therefore proceed to consider how the mean resistance, the profile of the mean velocity and the mean concentration of suspended matter in the pipe are affected by turbulence.

Resistance. In laminar flow, such as the motion of a viscous liquid in a capillary tube, observations confirm the well-known theoretical solution of Poiseuille (1840). This shows that the pressure drop over a given length of tube is directly proportional to the discharge or the quantity of fluid passing through the tube in unit time. In turbulent flow the fall of pressure is more nearly proportional to the square of the discharge. This change is typical of resistance laws; for small Reynolds numbers ($Re<1$) the resistance experienced by a sphere in translation through a viscous fluid varies directly as the velocity (Stokes 1851), but at higher Reynolds numbers the variation is approximately as the square of the speed, the change being essentially due to the formation of a turbulent wake.

Velocity Profile. Of more direct consequence in the study of atmospheric turbulence is the change which takes place in the variation of the mean velocity across the pipe as turbulence sets in. In Poiseuille flow the velocity profile is

$$u(\zeta)=\frac{1}{4\mu}\left(\frac{p_1-p_2}{l}\right)(r^2-\zeta^2) \qquad . \qquad . \qquad 1.31$$

where $p_1 - p_2$ is the pressure drop over the length l, r is the radius of the pipe and ζ is the (variable) distance from the axis of the pipe. The profile is thus parabolic from the axis to the wall and the velocity gradient is a linear function of distance from the pipe axis.

In turbulent flow the profile of mean velocity is entirely different; it is much more uniform over the central core and decreases sharply to zero on approaching the wall (Fig. 1).

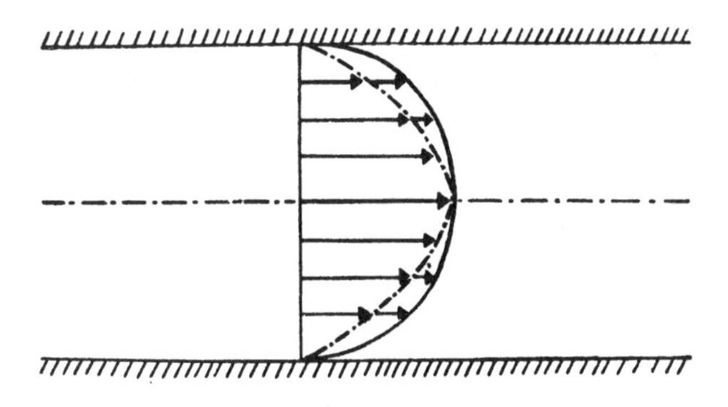

FIG. 1.—Velocity profiles in pipe flow

—·—·—· laminar flow
——— turbulent flow

(Since there is no exact theory available for turbulent motion even in the relatively simple case of pipe flow there is no exact theoretical formula which may be contrasted with 1.31 above. Semi-empirical formulae have been devised for profiles in turbulent flow and these are quite unlike the expressions met with in laminar flow since they usually involve small fractional powers of the distance from the boundary (e.g. $z^{1/7}$) or transcendental functions (e.g. $\log z$). The mean velocity gradient in turbulent pipe flow is thus small everywhere in the fluid except very near the wall, where it attains large values.

These observations are of the greatest importance in the

study of turbulence and particularly so in the atmosphere, where an analogous state of affairs is found. As will be seen later, the gradient of mean velocity is a useful indicator of turbulence in certain circumstances.

Diffusion of Mass. In Reynolds' experiment the presence of turbulence is shown vividly by the break-up of the filament of dye and the rapid spreading of the colour throughout the whole fluid. The effect of the turbulence has been to replace the slow molecular diffusion by a much more speedy process acting on a larger scale, and this again is typical of the turbulent state.

The Definition of Turbulence. So far we have avoided defining turbulence except in general qualitative terms. There is, in fact, no exact definition of turbulent flow in the sense in which, for example, irrotational flow is defined in the classical hydrodynamics. *Physically*, the turbulent state is unmistakable and easily described in general terms; *mathematically*, probably the most that can be said at present is that turbulence implies a state in which the instantaneous velocities exhibit irregular and apparently random fluctuations so that in practice only statistical properties can be recognized and subjected to analysis. The situation is, in fact, analogous to that accepted unreservedly in the field of molecular physics—exact knowledge of the motion of individual molecules either cannot be obtained, or if obtained is too complex to be of use, but given an assembly of molecules it is possible to deduce the laws of the behaviour of gases as statistical truths. A similar philosophy has shaped the theory of turbulence.

4. MATHEMATICAL APPROACH TO THE THEORY OF TURBULENT MOTION

Equations of Motion. If u, v and w are the component velocities of a viscous incompressible fluid along axes of x, y and z respectively, the equations of motion in the absence of external forces (such as gravity) have the form:

$$\rho\left(\frac{\partial u}{\partial t}+u\frac{\partial u}{\partial x}+v\frac{\partial u}{\partial y}+w\frac{\partial u}{\partial z}\right)=-\frac{\partial p}{\partial x}+\mu\left(\frac{\partial^2 u}{\partial x^2}+\frac{\partial^2 u}{\partial y^2}+\frac{\partial^2 u}{\partial z^2}\right), \quad . \quad 1.41$$

(with two equations of the same form for v and w), where $\partial p/\partial x$, &c., are the pressure gradients along the three axes and ρ is the density of the fluid.* These equations, since they involve the squares and products of the unknown velocities, are non-linear and therefore, in general, present insuperable difficulties to the mathematician, who is no longer able to make use of standard linear equation technique (such as the principle of superposition of solutions) for their resolution. Exact solutions of the equations are known only for a few limiting cases of low Reynolds numbers and simple geometrical configuration (e.g. Stokes' investigation for the sphere which omits the non-linear inertia terms), and it may be accepted that any approach to the problem of turbulence which proceeds via a frontal attack on the complete equations of motion is unlikely to succeed.

We may, however, elucidate the problem considerably by the introduction of mean values after the manner of Reynolds.

Mean Values. The *mean value* of the velocities u, v, w over an interval T is defined as

$$\bar{u}=\frac{1}{T}\int_{t-\frac{1}{2}T}^{t+\frac{1}{2}T} u\,dt, \text{ \&c.}$$

A *fluctuation* or *eddy velocity* (u') is defined as the difference between an instantaneous velocity (u) and a mean velocity ($\bar{u}$) so that

$$u=\bar{u}+u' \qquad v=\bar{v}+v' \qquad w=\bar{w}+w'.$$

If the interval T is taken long enough to include a large number of oscillations we have

$$\bar{u}'=\bar{v}'=\bar{w}'=0.$$

The motion thus consists of a mean flow (which in our problems is usually steady) on which is superimposed a velocity oscillating irregularly above and below the mean so that over a sufficiently long period of time the sum of the positive fluctuations equals the sum of the negative fluctua-

* These equations are now generally called the 'Navier-Stokes equations of motion'.

tions and therefore, on taking the average, cancel. It should be noted that the mean values of the squares or products of the eddy velocities are not necessarily zero.

The Reynolds Stresses. The effect of viscosity is to produce a system of stresses in the fluid, such as the shearing stress considered in our discussion of Couette flow above. In the case we are now considering the component stresses are six in number, p_{xx}, p_{yy}, p_{zz}, p_{yz}, p_{zx}, p_{xy}, defined as follows for an incompressible fluid.

$$p_{xx}=-p+2\mu\ \partial u/\partial x$$

$$p_{yz}=\mu(\partial w/\partial y+\partial v/\partial z),$$

together with four similar equations for the remaining stresses.*

The equations of motion then take the form

$$\rho\frac{\partial u}{\partial t}=\frac{\partial}{\partial x}(p_{xx}-\rho u^2)+\frac{\partial}{\partial y}(p_{xy}-\rho uv)+\frac{\partial}{\partial z}(p_{xz}-\rho uw), \text{ \&c.} \quad . \quad 1.42$$

Substituting $\bar{u}+u'$ for u, &c., and taking means we find

$$\rho\frac{\partial \bar{u}}{\partial t}=\frac{\partial}{\partial x}(\bar{p}_{xx}-\rho\bar{u}^2-\overline{\rho u'^2})+\frac{\partial}{\partial y}(\bar{p}_{yx}-\rho\bar{u}\bar{v}-\overline{\rho u'v'})$$

$$+\frac{\partial}{\partial z}(\bar{p}_{zx}-\rho\bar{u}\bar{w}-\overline{\rho u'w'}), \text{ \&c.} \quad . \quad 1.43$$

The set of equations 1.43 is identical in form with 1.42 above if we replace the stress p_{xx} by $\bar{p}_{xx}-\overline{\rho u'^2}$, p_{xy} by $\bar{p}_{xy}-\overline{\rho u'v'}$ and p_{zx} by $\bar{p}_{zx}-\overline{\rho u'w'}$. The only change is therefore the addition to the original viscous stresses of certain terms depending on the eddy velocities. The additional terms are called the *Reynolds stresses* and represent the effect of the fluctuations on the transport of momentum across a surface in the fluid.

These virtual stresses are invariably much greater than the corresponding viscous stresses and it is usually possible to ignore the latter in problems of turbulent motion. We

* Lamb, *Hydrodynamics* (6th ed.), art. 326. Actually there are 9 component stresses, but in the absence of body forces these reduce to 6 independent stress components.

shall be mainly concerned with terms of the type $-\rho\overline{u'w'}$ and the physical significance of this type of term is best realized by an appeal to statistical theory. The *coefficient of correlation* between two variable quantities u' and w' is defined as

$$r=\frac{\overline{u'w'}}{\sqrt{(\overline{u'^2})}.\sqrt{(\overline{w'^2})}}.$$

If $\overline{u'w'}=0$, $r=0$ and there is no correlation between u' and w' and conversely $r=0$ implies that $\overline{u'w'}=0$, so that the value of $-\rho\overline{u'w'}$ depends not only upon the magnitudes of u' and w' but also upon the degree of correlation between these two velocities. How such a correlation may arise may be seen by considering the important case in which u represents the component of velocity in the direction of the mean wind near the ground and w the velocity in the vertical direction. Since horizontal gusts (positive values of u') are more often associated with momentary downward currents (negative w') than with upward currents, and lulls (negative u') are more often associated with upward currents (positive w') than with downward currents, it is clear that in general a correlation will exist between u' and w' and the corresponding Reynolds stress $-\rho\overline{u'w'}$ will have a non-zero value. On the other hand, variations in w' are not systematically associated with particular lateral eddy velocities v' so that in this case the corresponding Reynolds stress $-\rho\overline{v'w'}$ is small or zero. These matters will be considered in greater detail later.

5. THE ANALOGY BETWEEN TURBULENCE AND MOLECULAR STRUCTURE

From the above considerations it becomes evident that there is a close analogy between the action of turbulence and the behaviour of an assembly of molecules, and most of the development of the theory has proceeded on these lines. In general language the *eddy*, regarded as a mass of fluid capable of retaining some measure of individuality and of moving as a separate entity in the surrounding fluid, takes the place of the molecule. In meteorology, if not so much in aerodynamics, it has become customary to speak of *eddy viscosity*, *eddy conductivity* and *eddy diffusivity* as the macroscopic counterparts of the molecular viscosity, conductivity and diffusivity. The analogy, however, cannot be carried too far and the difference is much more than one of scale. It is this fact which gives rise to the difficulties which will be discussed in the succeeding chapters.

REFERENCES

For a more detailed account of the subject-matter of this chapter the reader should consult

Aerodynamic Theory, ed. W. F. Durand, Vol. III, Section G (1934)

Applied Hydro- and Aeromechanics, by L. Prandtl and O. G. Tietjens, translated by J. P. Den Hartog (1934)

CHAPTER II

THE METEOROLOGY OF THE LOWER ATMOSPHERE

IN this chapter we consider the special properties of the lower atmosphere which call for explanation. In general, the term 'lower atmosphere' will be used to denote the layers of air extending from the ground to heights not exceeding about 100 m.

1. GUSTINESS AND THE DIURNAL VARIATION OF TURBULENCE

The most direct and easily recognized evidence of turbulence is the record of an anemometer which is capable of indicating instantaneous values of the wind speed, such as the Dines pressure-tube instrument. A typical trace obtained at a height of about 13 m. over downland in clear summer weather is shown in Fig. 2. The outstanding features are that the velocity exhibits rapid fluctuations over most of the period, and that the amplitude of the oscillations changes considerably throughout the 24 hours, tending to a maximum in the midday period and falling to a low value during the night. This manifestation of turbulence is usually referred to as *gustiness*. Fluctuations of this type also occur in the direction of the wind, so that it is possible, with suitable instrumental aids, to evaluate not only the mean wind speed over a period but also the magnitude of the component oscillations referred to any convenient system of axes.

The system of axes usually adopted is: x-axis along the direction of the mean wind (assumed constant over the period in question), y-axis across wind and z-axis vertical. A convenient numerical measure of the gustiness is the root-mean-square value of the ratio of the oscillations to the

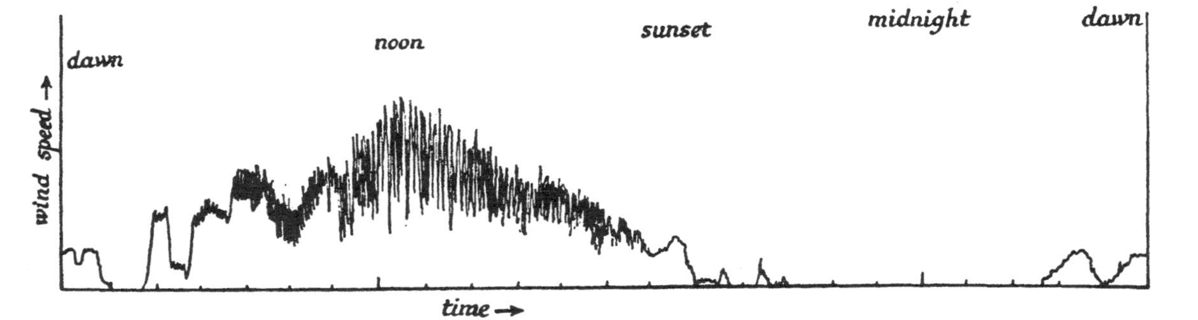

FIG. 2.—Typical record of wind speed (pressure-tube anemometer) at a height of about 13 m. over downland in 24 hours of clear weather

mean wind. In the notation of the previous chapter we have the scheme below which will usually be followed throughout the book.

Orientation of co-ordinate axis	Co-ordinate	Component of mean wind	Fluctuation (eddy velocity)	Gustiness components
In direction of mean wind .	x	$\bar{u}$	u'	$\sqrt{(\overline{u'^2}/\bar{u}^2)}=g_x$
Across mean wind . .	y	0	v'	$\sqrt{(\overline{v'^2}/\bar{u}^2)}=g_y$
Vertical . . .	z	0	w'	$\sqrt{(\overline{w'^2}/\bar{u}^2)}=g_z$

The detailed studies of Taylor, Scrase and Best, which will be referred to later, have established that in the first few metres above the ground $g_x \neq g_y \neq g_z$, but all three components of gustiness are of the same order of magnitude and show the same type of diurnal variation in clear weather.

The physical reasons for the diurnal variation of gustiness become clearer when records for overcast windy days and nights are compared with those for clear skies. When considerable cloud is present the fluctuations remain in sensibly constant ratio to the mean wind throughout the whole period, i.e. gustiness is much the same day and night. This implies that the diurnal variation of turbulence is connected with variations in the temperature of the ground, since in overcast weather the cloud cover tends to prevent radiation from reaching or leaving the surface in large amounts. This in turn must affect the gradient of temperature in the lower layers of air, and so we are led to seek a possible close correlation between the rate of change of temperature with height and the degree of turbulence shown by the wind.

2. THE DIURNAL VARIATION OF TEMPERATURE GRADIENT

The systematic recording and study of the vertical gradient of temperature near the ground (from 2·5 cm. to

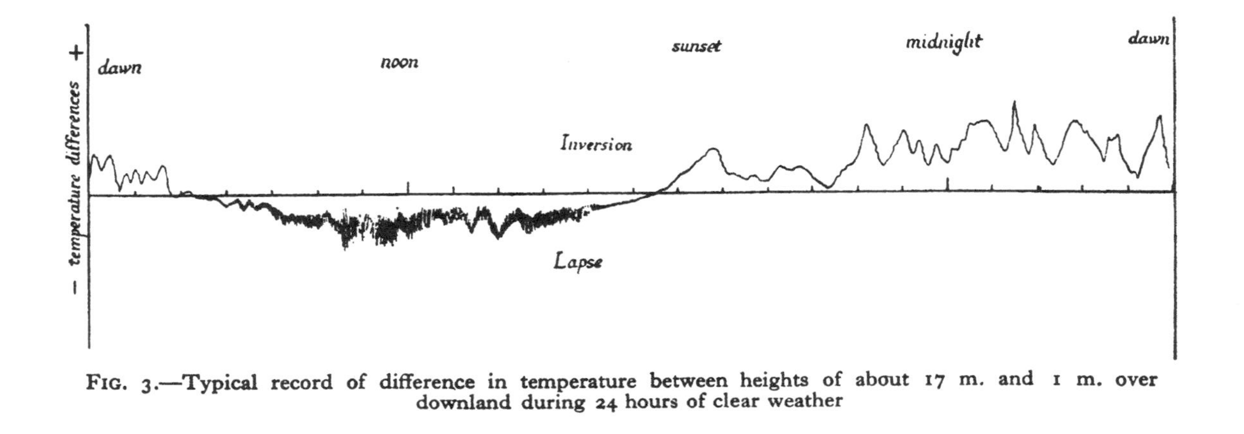

FIG. 3.—Typical record of difference in temperature between heights of about 17 m. and 1 m. over downland during 24 hours of clear weather

17·1 m.) has been carried out at the Experimental Station, Porton, Wiltshire, since 1921 and the main results are given in the Geophysical Memoirs of Johnson (No. 46) and Best (No. 65). A typical record of the difference in temperature between heights of 1·2 m. and 17·1 m. during a period of 24 hours of clear weather is shown in Fig. 3, which indicates a diurnal variation of the same type as that found for gustiness. Soon after dawn in clear weather the incoming radiation starts to raise the temperature of the ground and the air at the lower levels becomes warmer than that at greater heights. As the sun ascends the rate of fall of temperature with height, or *lapse rate*, increases in magnitude and attains a maximum value in the hours around noon. As the sun descends the lapse rate decreases, until at about an hour before sunset an approximately isothermal state is reached in the lower atmosphere. After sunset the ground loses heat rapidly because of the escape of long-wave radiation to space; the air nearest the ground becomes colder than the air above and the so-called *inversion* appears, in which temperature increases with height. Throughout the night the inversion usually shows large long period fluctuations, but in general persists until shortly after dawn when the whole cycle starts again. In the case of an overcast sky with a moderate or strong wind the temperature gradient remains very small throughout the entire period.

These observations indicate in a qualitative manner why turbulence shows a marked diurnal variation in clear weather. A large lapse rate implies that warmer and therefore less dense air lies below colder and therefore denser air. Such a disposition is favourable to the formation of convection currents, and any mass of air displaced slightly upwards is likely to continue to rise because of the difference in density between it and its surroundings. This is therefore a state of instability in which any disturbance, whether of mechanical or thermal origin, is likely to persist or even increase, a condition which is obviously favourable to the growth of turbulence. An inversion, on the other hand, means colder and therefore denser air lying below warmer and therefore less dense air, a condition of stability in which any disturbance tends to be damped out. Large lapse rates are therefore accompanied by vigorous oscillations of the wind and large inversions imply an approach to laminar flow.

The diurnal variation of turbulence is frequently seen without instrumental aids. The smoke from a weed fire in the country is scattered over a wide arc in the middle hours of a warm day, but on a clear night the same fire will give a thin dense plume or sheet of smoke which will drift for considerable distances without appreciably mixing with the surrounding air. This is Reynolds' classic experiment again but on a larger scale and with a new factor, the gradient of density, mainly controlling the transition from laminar to turbulent flow.

3. THE ADIABATIC LAPSE RATE AND ITS SIGNIFICANCE

A volume of air forced to ascend will expand because of the falling pressure and will therefore grow colder as it rises. This observation gives rise to the condition of vertical stability in the atmosphere; since the moving mass of air takes up instantaneously the pressure of its surroundings, its density relative to its environment will be determined by its absolute temperature.

It is shown in textbooks of dynamical meteorology* that an atmosphere in neutral vertical equilibrium, i.e. one in which an ascending or descending mass of air always has the same density as its environment, possesses a characteristic temperature distribution of the form

$$T(z)=T(0)-\Gamma z,$$

where $T(z)$ is the absolute temperature at height z above the surface ($z=0$) and Γ is the so-called *adiabatic lapse rate.* In dry air

$$\Gamma=gA/c_p=9{\cdot}86\times10^{-5}\ ^\circ\text{C./cm} \sim 1^\circ\ \text{C. per 100 m.}$$

where g is the acceleration due to gravity, A is the reciprocal of the mechanical equivalent of heat and c_p is the specific

* See, for example, Brunt, *Physical and Dynamical Meteorology* (Cambridge), Chapter II.

heat of air at constant pressure. This means that one result of thoroughly churning up the atmosphere would be to make the temperature gradient everywhere equal to the adiabatic lapse rate; this is in contrast to what happens with an incompressible fluid (for example, with a shallow bath of liquid) where the result of stirring is to produce an isothermal state, i.e. a zero lapse rate. This result may also be expressed by the statement that the effect of stirring the atmosphere is to produce a state in which entropy is invariable with height.

The adiabatic lapse rate is of fundamental importance in meteorology, since it represents the equilibrium state to which the atmosphere is always trying to return. Normally, the temperature gradient in the lower atmosphere varies considerably with height; Best (op. cit., p. 14) has found that in the layer 2·5–30 cm. above a close-cropped lawn the summer midday lapse rate is always large and can be nearly a thousand times the dry adiabatic rate, but such enormous values are necessarily confined to shallow layers very near the ground.* In the upper atmosphere, below the tropopause, the average gradient is about two-thirds of the dry adiabatic lapse rate. On overcast windy days and nights, when turbulence causes thorough mixing, the temperature gradient near the ground is invariably very close to the adiabatic lapse rate.

The significance of the adiabatic lapse rate in problems of atmospheric stability has led meteorologists to introduce the concept of *potential temperature*, defined as the temperature which a mass of air attains when brought adiabatically to a standard pressure. Thus an atmosphere possessing an adiabatic lapse rate is one in which potential temperature is invariable with height, and the atmosphere is in stable, unstable or neutral equilibrium according as the potential temperature increases, decreases or is constant with height.

4. THE VELOCITY PROFILE NEAR THE GROUND

In the section on pipe flow in Chapter I it was shown that the mean velocity profile near a rigid boundary is pro-

* For a discussion of this feature of the lower atmosphere see Brunt, *Physical and Dynamical Meteorology*, Chapter XII.

foundly affected by the type of flow. This is also true of the mean wind profile near the ground.

The simplest expression for the mean wind profile, both as regards analysis of the observations and the subsequent mathematical work, is undoubtedly a power law. This is usually written in the form

$$\bar{u}=\bar{u}_1(z/z_1)^p, \qquad p\geqslant 0 \quad . \quad . \quad . \quad 2.41$$

where $\bar{u}$ is the mean wind at height z and $\bar{u}_1$ is the mean wind speed at the constant reference height z_1. This equation implies that $\bar{u}=0$ on $z=0$ and also that $\bar{u}\rightarrow\infty$ as $z\rightarrow\infty$, so that the equation can at the most be applicable only to relatively shallow layers near the ground.

Recently, very detailed and carefully executed examinations of wind profiles over level short grass surfaces have shown that in conditions of neutral equilibrium (temperature gradient very close to the adiabatic value), the $\bar{u}$, $\log z$ plot is accurately linear for $0<z\leqslant 10$ m. In large lapse rates and large inversions there are marked systematic departures from linearity, i.e. $\bar{u}$ is no longer strictly proportional to the logarithm of the height. The power-law representation is to be regarded as a convenient interpolation formula for conditions near the surface. For profiles in the much deeper layer 1·5–120 m., R. Frost (1947) finds that a power law is fairly well followed over a wide range of conditions.

Of more immediate interest, however, is the question of the type of variation which the velocity profile shows with temperature gradient. The results of a long series of observations made by Giblett at Cardington (1932) indicate that the index of z in equation 2.41 is closely correlated with the magnitude and sign of the temperature gradient, with p ranging from about 0·01 in high lapse rates to 0·62 in a large inversion. A similar variation but with p ranging from 0·145 to 0·77 has been given by Frost (loc. cit.) for 1·5 m.$<z<$120 m. The value p=0·13 was found by Scrase (op. cit.) to be appropriate to the velocity profile from 3 m. to 13 m. over downland in conditions of small temperature gradient.

The Effects of 'Smooth' and 'Rough' Boundaries. Scrase's result quoted above is of considerable interest in that it suggests a well-known aerodynamic profile. For turbulent flow over smooth boundaries at relatively high Reynolds numbers wind-tunnel investigations indicate a profile of the form $\bar{u}=\bar{u}_1(z/z_1)^{1/7}$ (usually known as the 'seventh-root profile') near the boundary. At higher Reynolds numbers the index decreases slightly, values of one-eighth or one-ninth being more appropriate.

When a surface is 'aerodynamically rough', i.e. when the boundary layer is turbulent right down to the surface, it is customary to introduce a parameter called the 'roughness length' which is supposed to be related to the average size of the obstacles (such as blades of grass) which cover the surface.* Investigations by Prandtl, Rossby and Montgomery, Sverdrup and others, and more recently by E. L. Deacon at Porton, indicate that for adiabatic lapse rates the wind structure in the lower atmosphere agrees with the aerodynamic equation

$$\bar{u}=(1/k)(\tau_0/\rho)^{1/2}\log(z/z_0,) \qquad (z \geqslant z_0) \quad . \quad 2.42$$

where k is a non-dimensional quantity known as Karman's constant and whose value in aerodynamical work is usually taken to be about 0·4, τ_0 is the shearing stress at the surface, and z_0 is the roughness length. This relation has been established by Nikuradse, Schlichting and others in the laboratory, and it is a matter of considerable interest to find that it also holds for flow on the atmospheric scale. Deacon has also found that for all temperature gradients the relation

$$d\bar{u}/dz=az^{-\beta} \quad . \quad . \quad . \quad . \quad . \quad 2.43$$

(where a is independent of height) holds, with $\beta>1$ for superadiabatic gradients, $\beta=1$ for adiabatic lapse rates, and $\beta<1$ for inversion conditions. These results imply the profile

$$\bar{u}=\frac{(\tau_0/\rho)^{1/2}}{k(1-\beta)}[(z/z_0)^{1-\beta}-1] \quad . \quad . \quad . \quad 2.44$$

* See Chapter IV for a detailed discussion of this matter.

for non-adiabatic gradients, and the logarithmic law 2.42 for adiabatic gradients.

The problem of roughness in general, and of the definition of the roughness length in particular, will be referred to later (Chapter IV). For the present it suffices to indicate the physical implications of the change of velocity profile. In general terms the turbulence, by bringing down fresh masses of rapidly moving air, renews the momentum being lost as a result of the friction of the surface so that when turbulence is at a high level (superadiabatic gradients) inequalities of velocity are smoothed out and (except very near the surface) the gradient is small. In inversion conditions the general suppression of turbulence means that the supply of momentum from the higher layers is diminished and the velocity gradient increases. The limiting cases are, in the notation of equation 2.41

very turbulent air; $p=0$, $\bar{u}=\bar{u}_1$, $d\bar{u}/dz=0$

non-turbulent air; $p=1$, $\bar{u}=\bar{u}_1 z/z_1$, $d\bar{u}/dz=\text{constant}\neq 0$

Neither of these limits is reached in practice but both are sometimes approached fairly closely in high lapse rates and large inversions, respectively. The velocity profile or its gradient is thus a sensitive and useful indicator of the degree of turbulence present in the air, and as such has been used by Sutton in the treatment of the diffusion of smoke and gases near the ground (see Chapter V).

5. THE EDDY VELOCITIES NEAR THE GROUND

The magnitudes and distribution of the eddy velocities, u', v' and w', have been studied, among others, by Taylor, Scrase and Best. All three measured g_y and g_z by the device known as the bi-directional vane (i.e. a vane free to move simultaneously in the vertical and horizontal directions and whose movements are recorded by a pen on a drum), and in addition Best used a hot-wire anemometer to measure u'. Scrase also carried out an analysis of ciné records of the motion of a light vane and a swinging plate.

Scrase's investigations were limited to conditions of adiabatic lapse rate. He divided his results into those typical of large-scale turbulence (mean values taken over an interval of 1 hour), intermediate-scale turbulence (mean values over intervals of the order of a few minutes) and small-scale turbulence (mean values over intervals of the order of seconds). The existence of this classification brings to light one of the main difficulties in atmospheric turbulence, namely, that the oscillations themselves have periods varying from fractions of a second to many minutes. In normal wind-tunnel work special devices are used to ensure that the turbulence is of short period and as uniform as possible, but in the atmosphere no such simplification is possible. This feature of natural winds is of special importance in problems of diffusion (see Chapter V).

Scrase found that all three components are proportional to the wind speed, i.e. gustiness is independent of the wind velocity. (This result may not hold for very low winds.) For intermediate-scale turbulence he found that w'/v' had the average value of 0·7 at a height of 2 m., i.e. near the surface the lateral eddy velocity is about 50 per cent. greater than the vertical oscillation. For small-scale turbulence he gave the relation $u':v':w'=1{\cdot}0:1{\cdot}16:0{\cdot}75$. Thus near the ground, turbulence is non-isotropic, but Best (op. cit), while confirming the general character of Scrase's results at the lower levels, concludes that at heights in excess of about 25 m. all three components are probably equal.

Best also made considerable investigations on gustiness and found a complex variation with height, namely, that between 25 cm. and 5 m., lateral gustiness decreases slowly with height but vertical gustiness increases with height. The empirical laws obtained were $g_y \propto z^{-0{\cdot}06}$ and $g_z \propto z^{0{\cdot}175}$. For the eddy velocities themselves Best found that both the lateral and vertical components increase with height over the layer 25 cm.–5 m., and he gave the empirical laws $v' \propto z^{0{\cdot}34}$ and $w' \propto z^{0{\cdot}11}$ for this interval.

It will thus be seen that the oscillations show a complicated structure near the ground and considerably more work, with sensitive instruments covering a greater range of heights, is needed before an adequate survey can be given. In a very general sense the oscillations must decrease with height since turbulence of the type found near the surface is not present in the upper atmosphere, but a complete picture of the variation with height does not exist at present.

The Frequency Distribution of Eddy Velocities. It has been shown by Hesselberg and Bjorkdal (1929) that the distribution of eddy velocities follows a law similar to that enunciated by Maxwell for molecular velocities. If $f(u')du'$ denotes the probability that the x-component lies between u' and $u'+du'$, the Maxwell law is

$$f(u')=\frac{k}{\sqrt{\pi}}\exp(-ku'^2).$$

This relation has been examined by Best (op. cit., p. 44) by plotting log f against g_x^2, where f is the number of fluctuations per unit range of g_x expressed as a percentage of the total examined. The result shows a very close fit to the law

$$f=f_0\exp(-kg_x^2),$$

where k is a dimensionless quantity of the order of 10^{-3} whose value changes slightly with the mean wind speed and the stability of the atmosphere.

6. SHEARING STRESS NEAR THE GROUND

The horizontal shearing stress τ due to turbulence is given by Reynolds' formula

$$\tau=-\rho\overline{u'w'}$$

(Chapter I). It is customary to treat τ as independent of height near the ground and equal to its value at the surface (τ_0), and the justification for this is to be found in papers by Ertel (1933) and Calder (1939). On this assumption Sheppard (1947) found the eddy shearing stress over a smooth concrete surface at Porton to lie between 0·2 and 0·3 g.cm.$^{-1}$sec.$^{-2}$ for winds at 2 m. between 3·7 and 5 m./sec. This result is probably the most reliable yet obtained but refers, of course, to a rather unnatural surface.

Sheppard obtained his estimate of τ_0 by measuring the

drag of the wind on a small horizontal test surface at ground level. Scrase (op. cit.) evaluated $-\rho\overline{u'w'}$ directly from the cine records of the movements of light vanes and found τ to be about 0·9 g.cm.$^{-1}$sec.$^{-2}$ at 1·5 m. but to increase to 3·62 g.cm.$^{-1}$sec.$^{-2}$ at 19 m. over downland, i.e. a fourfold increase of τ with height over the interval 1·5–19 m. Thus the only direct observation on the variation of eddy shearing stress with height near the ground contradicts the assumption, so often made, that τ is virtually independent of height in the lower layers of the atmosphere. As we shall see later, this assumption is fundamental in most, if not all, theories of atmospheric turbulence and the difficulty is at present unresolved. There is obviously a need for more experimental work here.

7. THE THERMAL STRUCTURE OF THE LOWER ATMOSPHERE IN CLEAR SUMMER WEATHER

The Flux of Heat. In the early hours of a clear summer morning there is a rapid upward transfer of heat from the surface to the air (vide Johnson, op. cit.), but it has been shown by Sutton (1948) that in the hours around noon on such a day conditions become remarkably steady, $\partial T/\partial t$ being small and approximately constant with height and time for heights between 7 m. and 45 m. Sutton found that in these conditions the upward flux of heat (q) is given by

$$q(z)=q(0)-5{\cdot}4\times10^{-8}z \text{ g.cal.cm.}^{-2}\text{sec.}^{-1} \qquad (z<100 \text{ m.})$$

where $q(0)$ is the value of q at the surface and z is measured in centimetres. This equation was deduced from observations made at Leafield, Oxfordshire, and is probably fairly representative of conditions over undulating cultivated land. The magnitude of q_0 was estimated to be between 2×10^{-3} and 5×10^{-3} g.cal.cm.$^{-2}$sec.$^{-1}$, so that the effect of the second term is not appreciable below 100 m. In other words during the mid-hours of a clear summer day the upward flux of heat near the surface is approximately constant with an average value of about 3×10^{-3} g.cal.cm.$^{-2}$sec.$^{-1}$ over grass in southern England.

Variation of Temperature Gradient with Height. Since the adiabatic gradient represents the equilibrium state for the atmosphere the relevant entity in discussing heat transfer is not the actual gradient of mean temperature but the difference between the observed gradient and the adiabatic lapse rate. For the mid-hours of a clear summer day Sutton (loc. cit.) has shown from the Leafield data that the relation

$$\frac{\partial \bar{T}}{\partial z}+\Gamma=\text{const. } z^{-1\cdot 75},$$

holds for heights between 7 m. and 45 m. This empirical law should not be taken to apply to the lowest layers, i.e. below 7 m.

A review of the broad features of the temperature field over land and sea has recently been given by P. A. Sheppard (1947), who finds that during the daylight hours temperature decreases roughly proportional to the logarithm of the height. Sheppard's analysis includes a survey of both the seasonal and diurnal variations of temperature gradient and contains much information which cannot be included here for reasons of space.

Temperature Fluctuations. The effect of turbulence is to produce fluctuations of temperature very similar to those found in the wind speed. Reliable data regarding these oscillations are very scarce, chiefly because the instruments used for recording temperatures near the ground are usually designed to yield mean values rather than to record faithfully short-period changes. Some observations for clear summer days have been given by Johnson and Heywood (op. cit.), and from these Sutton (1948) has shown that for heights between 7 m. and 45 m. the amplitude of the fluctuations, as recorded, follows the empirical law

$$|T'|=\text{const. } z^{-0\cdot 4} \qquad (7 \text{ m.}<z<45 \text{ m.})$$

Some measurements of T' have also been given by Barkow (1915).

8. THE GRADIENT OF HUMIDITY IN THE LOWER ATMOSPHERE

The gradient of humidity is important for two reasons: in the first place it is clearly one of the chief factors controlling evaporation, and secondly, it has recently been shown by Booker and others (1947) that because of its influence on the refractive index the humidity gradient is of importance in the study of the propagation of high-frequency electro-magnetic waves over the surface of the earth.

Existing information on humidity gradient has been well summarized by Sheppard (loc. cit., 1947). The general features are as follows: in lapse conditions vapour pressure decreases with height even when the ground appears dry. In the inversion period the vapour-pressure gradient frequently decreases to zero and becomes positive (vapour pressure increasing with height) with deposition of dew, but if the air mass is sufficiently dry a weak negative gradient may persist throughout the night. Sheppard has also shown that over the sea the vapour pressure near the surface is approximately proportional to the logarithm of the height.

REFERENCES

Johnson, N. K. 1929. Meteorological Office, Geophysical Memoir No. 46

Best, A. C. 1935. Meteorological Office, Geophysical Memoir No. 65

Frost, R. 1947. *Meteorological Magazine*, 76

Giblett, M. A., *et al.* 1932. Meteorological Office, Geophysical Memoir No. 54

Scrase, F. J. 1930. Meteorological Office, Geophysical Memoir No. 52

Hesselberg and Bjorkdal. 1929. *Beit. Phys. fr. Atm.*, 15

Ertel, H. 1933. *Met. Zeit.*, 50

Calder, K. 1939. *Q.J. Roy. Meteor. Soc.*, 65

Sheppard, P. A. 1947. *Proc. Roy. Soc.*, A, 188

Sheppard, P. A., and Booker, H. G. 1947. *Meteorological Factors in Radio Wave Propagation* (published by The Physical Society)

Sutton, O. G. 1948. *Q.J. Roy. Meteor. Soc.*

Barkow, E. 1915. *Met. Zeit.*, 32

CHAPTER III

EARLY THEORIES AND THEIR APPLICATION TO LARGE-SCALE PHENOMENA

IN Chapter I it was pointed out that the problem of viscosity may be approached either by considering the bulk properties of a real fluid and postulating a system of forces to account for them, or by assuming a molecular structure and deducing viscous effects as consequence of this structure. In this chapter we shall be concerned with the counterpart in turbulence theory of the first of these approaches, primarily due to Taylor and Richardson in this country and to Wilhelm Schmidt in Austria, with whom the systematic study of atmospheric turbulence may be said to have originated.

1. WIND IN THE FRICTION LAYER

At heights well above the disturbing effects of the earth's surface the motion of the air conforms to a balance between the pressure gradient and the forces which arise because the air is moving on a rotating earth, and, in general, on a curved path. When the mean motion is steady and the isobars are straight or only slightly curved the balance is expressed by the simultaneous equations

$$\left.\begin{aligned} -2\omega \sin\phi\,\bar{v} &= -\frac{1}{\rho}\frac{\partial p}{\partial x} \\ 2\omega \sin\phi\,\bar{u} &= -\frac{1}{\rho}\frac{\partial p}{\partial y} \end{aligned}\right\} \quad . \quad . \quad . \quad . \quad . \quad 3.11$$

where ω is the angular velocity of rotation of the earth, ϕ is the latitude, and $\bar{u}$ and $\bar{v}$ are components of the mean velocity along axes pointing east and north respectively. The wind which satisfies the above equations is called the *geostrophic wind* and it is thus assumed to be steady, horizontal, and free from friction.*

In moderate or high latitudes the geostrophic wind is a useful approximation to the actual wind at about 500 m. above the surface. The problem which will now be discussed is how the geostrophic wind is modified by eddy friction as the surface is approached.

Following the analogy with viscosity we express the effect of friction by introducing virtual stresses τ_{zx} and τ_{zy} without inquiring, at this stage, how they arise. Writing $\lambda=2\omega \sin\phi$ and neglecting any variation of the stresses in the horizontal we have, inside the friction layer (0–500 m.)

$$\left.\begin{aligned} -\lambda\bar{v} &= -\frac{1}{\rho}\frac{\partial p}{\partial x}+\frac{1}{\rho}\frac{\partial}{\partial z}\tau_{zx} \\ \lambda\bar{u} &= -\frac{1}{\rho}\frac{\partial p}{\partial y}+\frac{1}{\rho}\frac{\partial}{\partial z}\tau_{zy} \end{aligned}\right\} \quad . \quad . \quad . \quad . \quad 3.12$$

These equations are, so far, exact for steady flow and straight isobars; we now take the analogy with viscosity a step further and write

$$\tau_{zx}=A_x\partial\bar{u}/\partial z; \qquad \tau_{zy}=A_y\partial\bar{v}/\partial z \quad . \quad . \quad 3.13$$

thus defining two coefficients A_x and A_y, analogous to the ordinary dynamic viscosity (μ), called by Schmidt *interchange coefficients* (*Austauschkoeffizient*). At this stage we have no information regarding the nature of these coefficients, which may thus be constants or functions of position, in this case, height above the surface. In the earliest treatment of the problem, due to Ekman and Taylor, the approximation $A_x=A_y=A=$constant was used. Taylor has been followed by most English writers in using not A but K, defined by $K=A/\rho$. K, termed the *eddy viscosity*, thus corresponds to the kinematic viscosity, and is usually expressed in cm.^{2}sec.$^{-1}$.

For convenience, the x-axis is taken along the isobar so that $\partial p/\partial x=0$, the pressure gradient being assumed invariable with height in the friction layer, and the components

* See Brunt, *Physical and Dynamical Meteorology*, Chapter IX.

of velocity are combined to form the vector $\bar{V}=\bar{u}+i\bar{v}$ where $i=\sqrt{-1}$. The total pressure gradient $\partial p/\partial y$ is replaced by $-\lambda\rho G$, where G is the geostrophic wind, so that neglecting the change of density with height equations 3.12 and 3.13 become

$$i\lambda(\bar{V}-G)-K\partial^2\bar{V}/\partial z^2=0,$$

to be solved with the boundary conditions

$$\bar{V}=0 \text{ on } z=0 \text{ (no slip at the surface).}$$

$$V\rightarrow G \text{ as } z\rightarrow\infty.$$

The details of the solution are discussed in textbooks of dynamical meteorology*; we find without difficulty

$$\begin{aligned}\bar{V}&=\bar{u}+i\bar{v}\\&=G[1-e^{-z\sqrt{(\lambda/2K)}}(\cos z\sqrt{(\lambda/2K)}-i\sin z\sqrt{(\lambda/2K)})]\quad . \quad 3.14\end{aligned}$$

This solution is usually displayed as a graph of $\bar{u}$ against $\bar{v}$, the curve being an equiangular spiral with the geostrophic wind as limit point (Ekman spiral). The solution also shows that inside the friction layer the wind no longer blows exactly along the isobars but is inclined towards the centre of low pressure. The angle between the wind and the isobar decreases with height and, since from eq. 3.14 $\lim_{z\to 0} \bar{u}/\bar{v}=1$, has a maximum value of $\pi/4$ at the surface. It may also be seen that if H is the height at which the wind attains the geostrophic direction,

$$H=\pi\sqrt{(2K/\lambda)}.$$

This relation enables an estimate to be made of the order of magnitude of K which until now has been brought in as a purely formal expression of the virtual frictional stresses. From Dobson's observations of pilot balloon trajectories over Salisbury Plain, Taylor found that K must be of the order of 10^4 cm.^{2}sec.$^{-1}$, i.e. at least a hundred thousand times greater than the kinematic viscosity of the air. The

* See, for example, Berry, Bollay and Beers, *Handbook of Meteorology* (1945), p. 453. A treatment with a somewhat different lower boundary condition, viz. that the direction of slip is that of strain, is to be found in Brunt, *Physical and Dynamical Meteorology*, Chapter XII.

method, of course, yields no information on the validity of the approximation K=constant.

Although it is now known that K cannot be constant with height, as a first approximation the wind spiral gives a fairly satisfactory representation of the effect of friction on the geostrophic wind throughout the deep frictional layer 0–500 m. The solution, however, departs radically from the truth near the ground.

2. THE DIURNAL TEMPERATURE WAVE AND ITS VARIATION WITH HEIGHT

An equally important large-scale problem is that of the variation with height of the diurnal temperature wave. For clear days the temperature of the surface of the earth may be approximately represented by a sine wave of period 24 hours, i.e. on $z=0$,

$$T=T_0+A\sin qt \quad . \quad . \quad . \quad . \quad . \quad 3.21$$

where $q=2\pi/24.60.60=7{\cdot}3\times10^{-5}$ and t is measured in seconds. The problem to be discussed is as follows: Given the boundary condition 3.21 above, what is the form of the diurnal wave at any height?

The virtual stresses τ_{zx}, τ_{zy} which were introduced to account for the frictional effects of turbulence indicate a large-scale diffusion of momentum which must also imply a simultaneous corresponding macroscopic transfer of heat and matter. The mechanism of such transfer is clearly that of the movement of large masses of air in the vertical, but this at once introduces the complication that in a compressible fluid, such as the atmosphere, any such vertical motion involves a change of density of the eddy, since a volume of air will change its temperature adiabatically in moving from one level to another because of the decrease of pressure with height. This complication does not, of course, arise in the ordinary kinetic theory of conduction in gases.

The equation of heat transfer by turbulence in a compressible fluid has been given by Brunt in the form*

$$\frac{\partial \bar{T}}{\partial t}=\frac{\partial}{\partial z}\left\{K\left(\frac{\partial \bar{T}}{\partial z}+\Gamma\right)\right\}, \qquad 3.22$$

where K is a macroscopic coefficient of conduction, called the *eddy conductivity*, not necessarily to be identified at this stage with the K used in the friction layer problem and which may be a constant or a function of the independent variables. Ignoring any possible effects of radiation, the flow of heat in the vertical thus depends on the difference between the actual gradient of mean temperature and the adiabatic lapse rate (Γ) and, in particular, the net flow is zero in an atmosphere which has attained the adiabatic temperature distribution. If the approximation K=constant is employed, equation 3.22 reduces to

$$\partial \bar{T}/\partial t=K\,\partial^2 \bar{T}/\partial z^2 \qquad 3.23$$

—the familiar equation of conduction of heat in a solid.

The problem of the propagation of the diurnal wave of temperature from the surface into an atmosphere of known linear temperature distribution for K=constant thus requires the solution of equation 3.23 with the boundary conditions

$$\left.\begin{array}{l}\bar{T}=\bar{T}_0+A\sin qt \text{ on } z=0\\ \bar{T}\rightarrow\bar{T}_0-\beta z \text{ for large } z\end{array}\right\} \qquad 3.24$$

where β is a constant. The required solution is given in textbooks of mathematical physics and is

$$T=T_0-\beta z+Ae^{-bz}\sin(qt-bz), \qquad 3.25$$

where $b^2=q/2K$. Comparison with observation can be made in two ways: (i) by considering the observed value of the ratio of the amplitudes of the diurnal wave at heights z_1 and z_2 and comparing it with the theoretical ratio $\exp b(z_2-z_1)$, or (ii) by evaluating the change of phase with height, a process most conveniently done by measuring the difference between the time of maximum temperature at heights z_1 and z_2, which can then be compared with the theoretical expression $\pi(z_1-z_2)b/2q$. These computations have been carried out by numerous workers, including Schmidt, Taylor,

* *Physical and Dynamical Meteorology*, Chapter XII.

Johnson and Best and their results* agree in ascribing to *K* values which are usually of the same order of magnitude as those found for the eddy viscosity in the problem of the friction layer. This suggests confirmation of what has been already implied in the notation, namely, that the two coefficients are identical.

A detailed examination of the results, however, shows that the expression 3·25 cannot be regarded even as a good first approximation to the actual observations, i.e. in this case the approximation K=constant is unacceptable. The application of the expression 3·25 to shallow layers of the atmosphere lying between 2·5 cm. and 87 m. has been made by Best (1935) and by Johnson and Heywood (1938), and their results show that the values of *K* deduced from both amplitude and phase-shift observations increase very rapidly with height (roughly as z^2 for clear summer days). Equation 3·25 also implies that the lag in the time of maximum temperature should increase linearly with the height, whereas Best found that the actual increase is very much slower—roughly as $z^{1/5}$—so that the solution obtained on the basis of K=constant is clearly of the wrong functional form.

There is nothing surprising in this: it is obvious that *K*, regarded as a quantity expressing the effects of turbulence in spreading heat through the atmosphere, would be expected to vary considerably during a period of 24 hours and over the considerable ranges of height involved if only because of the diurnal variation and change with height of turbulence itself. The statement sometimes found in meteorological texts that the turbulent transfer of heat in the lower atmosphere implies an eddy conductivity of the order of 10^3 to 10^5 cm.^{2}sec.$^{-1}$ has little meaning unless adequately qualified. To apply to any particular problem a value of *K* selected at random from the mass of results obtained by the use of equation 3·23 is extremely hazardous. From 2·5 cm. to 87 m.—the range of heights covered by Geophysical Memoirs No. 65 and 77—values of *K* ranging

* For an account of these results, see Brunt, *Physical and Dynamical Meteorology*, Chapter XII.

from a few tens to tens of thousands cm.^{2}sec.$^{-1}$ have been obtained, so that the application of K-values derived from one set of observations to any other circumstances may easily lead to results which are wrong in order of magnitude. In other words, the term $\frac{\partial K}{\partial z}\left(\frac{\partial \bar{T}}{\partial z}+\Gamma\right)$ cannot, in general, be neglected in equation 3.22.

3. THE DIFFUSION OF MATTER

Under this heading we shall consider two distinct types of problems: (i) those in which the source is independent of the characteristics of the ambient atmosphere and (ii) those in which the source strength depends upon the state (velocity, temperature, moisture content) of the air stream. The first category includes most sources of atmospheric pollution, such as factories, and in the second category the outstanding problem is that of evaporation.

Sources of Constant Strength. A quantity of matter in gaseous or finely divided form released into the atmosphere will diffuse under the combined influence of molecular and eddy forces. Introducing macroscopic diffusion coefficients, analogous to those considered above, the fundamental equation, neglecting molecular forces, is

$$\frac{\partial \bar{\chi}}{\partial t}+\bar{u}\frac{\partial \bar{\chi}}{\partial x}+\bar{v}\frac{\partial \bar{\chi}}{\partial y}+\bar{w}\frac{\partial \bar{\chi}}{\partial z}=\frac{\partial}{\partial x}\left(K_x\frac{\partial \bar{\chi}}{\partial x}\right)+\frac{\partial}{\partial y}\left(K_y\frac{\partial \bar{\chi}}{\partial y}\right)+\frac{\partial}{\partial z}\left(K_z\frac{\partial \bar{\chi}}{\partial z}\right), \quad 3.31$$

where $\bar{\chi}$ (g.cm.$^{-3}$) is the mean density of suspended matter; the introduction of three diffusion coefficients corresponding to the three co-ordinate axes is rendered necessary by the non-isotropic nature of turbulence near the ground. Taking the usual system of axes (Chapter II, p. 16) and assuming $K_x=K_y=K_z=K$=constant as a first approximation, equation 3.31 becomes

$$\frac{\partial \bar{\chi}}{\partial t}+\bar{u}\frac{\partial \bar{\chi}}{\partial x}=K\nabla^2\bar{\chi}. \quad . \quad . \quad . \quad . \quad 3.32$$

The most striking feature of the early work on atmospheric diffusion is that investigations employing solutions of equation 3.32 revealed an apparent increase of K with the distance from the source, i.e. with the scale of the phenomena. L. F. Richardson (1922) evaluated K in this way for the diffusion of smoke over short distances, for the distribution of volcanic ash and for the scatter of small free balloons, and found values ranging from 10^4 to 10^8 cm.^{2}sec.$^{-1}$. Even larger values (up to 10^{11} cm.^{2}sec.$^{-1}$) have been given by other writers.

The most important problems of diffusion are those associated with continuous sources, such as domestic or factory chimneys or military smoke generators. We shall here consider briefly only the steady continuous-point source, the solution of which for the approximations K=constant, $\bar{u}$=constant was given in 1923 by O. F. T. Roberts as

$$\bar{\chi}(x, y, z)=\frac{Q}{4\pi Kr}\exp\left\{-\frac{\bar{u}(y^2+z^2)}{4Kr}\right\}, \quad . \quad . \quad 3.33$$

where $r^2=x^2+y^2+z^2$ and Q is the strength of the source (g.sec.$^{-1}$) placed at the origin. This expression has been compared with accurate measurements of the diffusion of smoke and gas from controlled sources made at the Experimental Station, Porton. At 100 yards downwind of the source it was found that the observations implied values of K of the order of 10^3 cm.^{2}sec.$^{-1}$ in adiabatic gradient conditions, but at greater distances considerably higher values of K were required to make the expression 3.33 agree with the measured values.

A detailed comparison of the theoretical solution with the experimental data reveals the main cause of these discrepancies. The solution 3.33 implies that the central or peak concentration, namely, that found on the axis of the cloud ($y=z=0$), is independent of the mean wind speed and decreases inversely as the first power of the distance from the source. Actually, the experiments indicate that the peak concentration closely follows the law

$$\bar{\chi}(x, 0, 0)=\text{const.}/\bar{u}x^m,$$

where $1 < m < 2$, with the value $m = 1 \cdot 76$ in adiabatic gradient conditions (Sutton, 1947). Thus the actual decay of concentration downwind is much more rapid than that deduced on the assumption K=constant, an observation which explains clearly why the value of K apparently increases with distance from the source. The approximation K=constant is thus unacceptable in problems of large-scale atmospheric diffusion. This problem is considered in greater detail in Chapter V.

Sources of Variable Strength. Evaporation. The solution of equation 3.31 for the steady two-dimensional evaporation problem (wetted surface infinite across wind and of finite extent downwind) was given, for K=constant, $\bar{u}$-constant, by Jeffreys in 1918. In view of the inapplicability of the approximations this solution is now of limited interest. This work has now been generalized for variable K and is described in Chapter V. No practical solution has yet been found for the three-dimensional problem (area of any shape, finite across wind and downwind) even for the relatively simple case of K=constant, $\bar{u}$=constant. The problem is one of considerable technical difficulty.

4. EMPIRICAL FORMULAE FOR K AS A FUNCTION OF HEIGHT

The failure of the assumption K=constant to do more than indicate the general form of the desired solutions marks the essential difference (apart from that of scale) between eddy and molecular transport phenomena. It thus becomes necessary to consider the possibility of the interchange coefficients varying with position in the field, and the most natural assumption to make is that K is dependent upon distance from the main sources of turbulence, in this case the surface of the earth.

A considerable literature now exists on the possibility of accounting for the observed facts by taking K to increase with height. Usually K is assumed proportional to a constant power of the height, a representation peculiarly appropriate in these problems because, in aerodynamics, power laws have been found to give good representations of properties such as resistance, profile shape and boundary layer thickness over a wide range of conditions, and also because exact solutions of the relevant differential equations are often relatively easily found when K and $\bar{u}$ are expressed by powers of the independent spatial variable. At this stage such formulae are, of course, entirely empirical.

We consider here some typical applications of such expressions.

The Friction Layer Problem for K variable with Height. The problem of the approach to the geostrophic wind in the friction layer was investigated in 1925 by Prandtl and Tollmien, who utilized the general resemblance between the motion of the air near the earth's surface and pipe flow to deduce the functional relation between the eddy viscosity and the height. The basis of their work was the seventh-root velocity profile (*vide* Chapter II, p. 22) which, with the assumption of a constant shearing stress expressed by $K\,\partial\bar{u}/\partial z$, implies that $K \propto z^{6/7}$. Later, in 1933, Köhler obtained the solution for the general case $K \propto z^m, m>0$, in terms of Bessel functions. These investigations are too long to be given here, but one important result may be derived very simply, namely, the effect of changing stability on the angle between the surface wind and the geostrophic wind.

Equations 3.12 with the x-axis along the isobar may be written

$$\partial\tau_{zx}/\partial z = -\lambda\rho\bar{v}; \qquad \partial\tau_{zy}/\partial z = -\lambda\rho(G-\bar{u}). \qquad 3.41$$

Integrating with respect to z,

$$(\tau_{zx})_{z=0} = -\lambda\int_0^h \rho\bar{v}\,dz; \quad (\tau_{zy})_{z=0} = -\lambda\int_0^h \rho(G-\bar{u})\,dz, \quad 3.42$$

where h is the height at which the turbulent frictional stresses vanish. These equations are, as before, quite general for steady motion and straight isobars.

The resultant shearing stress at the surface is in the direction of the surface wind, so that if α be the angle between the surface wind and the geostrophic wind

$$\tan\alpha = (\tau_{zy}/\tau_{zx})_{z=0} = \lim_{z\to 0}(\bar{v}/\bar{u}). \quad . \quad . \quad 3.43$$

To evaluate this expression we introduce the assumption that in the friction layer the wind profile is represented by an empirical power law. (This is equivalent to assuming that K is proportional to a power of the height.) We write

$$\bar{u}=G(z/h)^p, \qquad p \geqslant 0,$$

so that $\bar{u}=0$ on $z=0$ and $\bar{u}=G$ on $z=h$. Since $\bar{v}=u \tan \alpha$ on $z=0$, and also $\bar{v}=0$ when $\bar{u}=G$ we have

$$\bar{v}=\bar{u}(1-z/h) \tan \alpha.$$

Substituting these expressions for $\bar{u}$ and $\bar{v}$ in equations 3.42 and 3.43 we find, without difficulty

$$\tan \alpha=\sqrt{p(p+2)}. \qquad . \quad . \quad . \quad . \quad . \qquad 3.44$$

This relation may be regarded as expressing the way in which the angle between the surface wind and the geostrophic wind changes with stability.

The limits of p are 0 and 1, the lower value indicating a completely turbulent atmosphere in which all inequalities of velocity are smoothed out, while the upper limit refers to laminar flow, but in practice, for all except the very lightest and consequently most erratic winds at the surface, the value of p only rarely exceeds $\frac{1}{2}$, while in conditions of adiabatic lapse rate we may take $p=\frac{1}{7}$. For this range of p we have the following values of α.

Angle between Surface Wind and Geostrophic Wind in Terms of the Velocity Profile

	'Stable' profiles			'Neutral' profile	'Unstable' profiles		
p	$\frac{1}{2}$	$\frac{1}{3}$	$\frac{1}{4}$	$\frac{1}{7}$	$\frac{1}{9}$	$\frac{1}{10}$	$\frac{1}{100}$
α	47°	41°	37°	29°	26°	24°	8°

It should be observed that the result previously given for K=constant, namely, that $\alpha=\pi/4$, is not obtained as $p \rightarrow 1$.

5. THE PROPAGATION OF THE DIURNAL TEMPERATURE WAVE WITH K A POWER OF THE HEIGHT

The solution of the equation

$$\frac{\partial T}{\partial t}=\frac{\partial}{\partial z}\left\{K_1 z^m\left(\frac{\partial T}{\partial z}+\Gamma\right)\right\} \quad . \quad . \quad . \quad . \quad 3.51$$

where K_1 is the value of K at unit height and $0\leqslant m<1$, subject to the boundary conditions 3.24, has been given by Köhler (1933) in terms of Bessel functions of imaginary argument. In this form the solution is not readily compared with observation, but by employing the asymptotic expansion for this class of Bessel functions it can be shown that for sufficiently large values of z the oscillatory term in the solution approaches the more manageable form:

$$T=\text{const. } z^{-\frac{1}{4}m}\exp\left(-\frac{2}{2-m}bz^{1-\frac{1}{2}m}\right).$$

$$\sin\left(2qt-\frac{2}{2-m}bz^{1-\frac{1}{2}m}\right) \quad . \quad 3.52$$

It follows from this expression that the time of maximum temperature should become later as the height increases, with the lag at the greater heights tending to the law

$$\text{lag}=\text{const. } z^{1-\frac{1}{2}m} \quad . \quad . \quad . \quad . \quad . \quad 3.53$$

It has been shown by Best (loc. cit.) that from the surface up to 17 m. the lag on a clear summer day satisfies the empirical law

$$\text{lag}=\text{const. } z^{0\cdot 19} \quad . \quad . \quad . \quad . \quad . \quad 3.54$$

and an examination of the data of Geophysical Memoir No. 77 indicates that the same law is valid up to about 90 m. in similar conditions. However, equation 3.53, being valid only for $m<1$, cannot be made to agree with the observations. The limiting case for which the solution holds, namely

$$K=K_1 z \quad . \quad . \quad . \quad . \quad . \quad . \quad 3.55$$

is inconsistent with the conclusion reached by many workers by applying the solution 3.24 for K=constant to a succession of shallow layers, namely, that the eddy conductivity must increase approximately as the square of the height in order to account for the transfer of heat in the vertical. The problem is considered at greater length in Chapter VI.

6. THE CONSTANCY OF THE SHEARING STRESS WITH HEIGHT AND THE CONJUGATE POWER LAW THEOREM

We conclude this chapter by proving a result of fundamental importance for wind structure near the ground. It may be easily shown by elementary arguments that for steady flow in pipes the shearing stress varies linearly with distance from the wall, a result equally valid for laminar or turbulent motion. In the case of flow adjacent to a plane boundary in the absence of a pressure gradient the shearing stress is necessarily constant and equal to its value at the surface (τ_0). For the atmosphere it has been shown by Ertel (1933) that the latter result is also valid (with an error not exceeding a few per cent) for heights up to about 25 m. at least.

Ertel's proof may be outlined as follows: taking axis fixed in the earth, we may write the gradient wind as the vector $Ge^{i\alpha}$ and similarly the wind at any level in the friction layer as $We^{i\psi}$ where α is the angle between the geostrophic wind and the x-axis and ψ is the corresponding angle for the wind in the friction layer. Taking $A_x=A_y=A$, a function of height, we can then write equations 3.12 as

$$\frac{\partial}{\partial z}\left\{A\frac{\partial}{\partial z}(We^{i\psi})\right\}-i\lambda We^{i\psi}+i\lambda Ge^{i\alpha}=0$$

The real part of this equation yields

$$\frac{\partial}{\partial z}\left(A\frac{\partial W}{\partial z}\right)-AW\left(\frac{\partial\psi}{\partial z}\right)^2-\lambda G\sin(\alpha-\psi)=0.$$

The change of mean wind direction with height is small over a shallow layer, so that we may neglect the term $AW(\partial\psi/\partial z)^2$ in comparison with the other two terms in the equation, and the condition that $\tau=A(\partial W/\partial z)$ shall be effectively constant from the surface to a height z becomes

$$\lambda W \int_0^z \sin(\alpha - \psi)\, dz \text{ to be very much less than } A(\partial W/\partial z).$$

Now A is of the order of 10 to 10^2 g.cm.$^{-1}$sec.$^{-1}$ at least, and by expressing W as a power of z Ertel shows that the above condition is satisfied for heights not exceeding about 25 m., and therefore that in this layer τ is constant (to within about 6 per cent).

Ertel's proof has been examined by K. L. Calder (1939), who pointed out that if the variation of wind direction with height be completely neglected, the mean motion in the lowest layers becomes strictly two-dimensional, and the theorem then reduces to the trivial case of W constant with height. Calder also showed that in spite of the extreme smallness of the change of the direction of the mean wind with height, a plausible assumption as to its order of magnitude in the lower layers suffices to account for the relatively large shear observed without in any way invalidating Ertel's final result.

The Conjugate Power-Law Theorem. Ertel's result leads to a relation of fundamental importance, originally conjectured by Schmidt. The theorem may be enunciated thus: *if the wind profile in the lower layers of the atmosphere can be represented by a power law $u=u_1(z/z_1)^p$, the eddy viscosity is given by $K=K_1(z/z_1)^{1-p}$ in the same layer.* This follows from equation 3.13 and the definition of K together with the proof that τ is constant near the ground.

7. GENERAL CONCLUSIONS ON THE '*K*-THEORY'

It is generally accepted by meteorologists that the development of the theory of turbulence on lines given in this chapter is now virtually complete and that little progress is to be expected by introducing further empirical refinements. The '*K*-theory' has been of immense service by clearing the path for more detailed investigations, and in particular the early work has established not only the salient features of the phenomena but also the orders of magnitude involved. It now seems clear that any particular series of observations, whether of the transport of momentum, heat or matter, can usually be fairly closely represented by the solution of the relevant equations with K proportional to a power of the height, but this type of analysis affords no clue to possible variations of the index with the stability of the atmosphere or with the nature of the surface. Considerations of this type indicate the necessity of more intimate studies of turbulence and an account of these forms the subject-matter of the next chapter.

REFERENCES

Schmidt, W. 1925. *Der Massenaustausch in freier Luft, Probleme der Kosmischen Physik*, 7

Taylor, G. I. 1915. *Phil. Trans. Roy. Soc.*, A, 215

Johnson, N. K., and Heywood, G. S. P. 1938. Meteorological Office, Geophysical Memoirs No. 77

Richardson, L. F. 1922. *Weather Prediction by Numerical Process*

Roberts, O. F. T. 1923. *Proc. Roy. Soc.*, A, 104

Jeffreys, H. 1918, *Phil. Mag.*, 35

Prandtl, L., and Tollmien, W. 1925. *Zeit. Geophys.*, 1

Köhler, H. 1933. *Meteorologische Turbulenzuntersuchungen*, K.S. Vet. Hang. (Stockholm), Bd. 13, Nr 1 (3rd series)

Ertel, H. 1933. *Met. Zeit.*, 50

Calder, K. L. 1939. *Q.J. Roy. Meteor. Soc.*, 65

CHAPTER IV

MIXING LENGTH AND STATISTICAL THEORIES OF TURBULENCE

IN this chapter we shall be concerned with the alternative approach to a general theory indicated in Chapter I, that is, with an examination of the intimate structure of the turbulent fluid itself, resembling in many respects the kinetic theory analysis of transport phenomena in gases. To watch, say, the diffusion of smoke near the ground suggests to the most casual observer that the growth of the cloud is caused by the apparently random motion of 'lumps' of air in all directions, each 'lump' carrying with it a content of suspended matter and thereby causing the macroscopic diffusion of the smoke. A similar process may be supposed to be going on with regard to momentum and heat so that in this respect the 'lumps' of fluid, or 'eddies', correspond to molecules in a gas.

A little reflection, however, soon shows that the analogy is by no means perfect. A satisfactory explanation of the broader aspects of the behaviour of gases can be derived from the simple hypothesis that the molecules are permanent bodies which behave like elastic spheres of fixed size. Further, this model is capable of considerable development without losing its essential mechanical character and has thus been extended to explain much of the detail of the behaviour of gases in respect of their viscosity, conductivity and diffusivity. The eddies in a turbulent fluid are primarily regions of concentrated vorticity and are thus evanescent and of variable 'size'. There is no immediately measurable property of turbulent flow which corresponds, for example, to the density of a gas (i.e. to the number and mass of molecules per unit volume), and instead of a definite dynamical event such as a collision between neighbouring molecules we can only speak, rather vaguely, of the eddy transferring a particular property by 'mixing' with the surrounding fluid.

There are at present two lines of attack prominent in the problem of the intimate structure of a turbulent fluid. The first, due to Prandtl (1925) and his fellow workers at Göttingen, attempts to set up a loose mechanical model of turbulence having obvious affinities with the Clerk Maxwell picture of a gas. The details of the model are never insisted upon, the whole scheme being used to promote the emergence of certain functions akin to those familiar in the kinetic theory and which are subsequently employed to facilitate the analysis of the complex phenomena observed in turbulent flow. The second and deeper approach, due primarily to Sir Geoffrey Taylor (1922, 1935), is philosophically more akin to the method of statistical mechanics in the analysis of molecular ensembles, the fundamental conception being that the properties of turbulent flow can be elucidated by a study of the correlation between velocities at neighbouring points or at successive instants of time. Both methods have been applied to meteorological problems.

1. THE CONCEPT OF THE MIXING LENGTH

We commence with an account of the essential ideas in Prandtl's treatment of the transfer of momentum by turbulence.

It has been shown in Chapter I, p. 11, equation 1.43, that the Reynolds stress corresponding to the term $\mu\, du/dz$ in laminar flow is

$$\tau = -\rho \overline{u'w'}.$$

In the case of a turbulent fluid in which the steady mean motion is entirely along the x-axis and is a function of z only, this equation represents the cross-current rate of transfer of momentum by the turbulence. The problem to be investigated is essentially that of expressing $\overline{u'w'}$ in terms of the mean velocity profile $\bar{u}(z)$ and its spatial derivatives $(d\bar{u}/dz, d^2\bar{u}/dz^2, \ldots)$. If this could be effected by exact analysis the main difficulty in turbulence theory would be resolved, but since so far this has proved impossible, recourse must be had to a simplified scheme in order to ascertain some plausible form for $\overline{u'w'}$ which will facilitate the analysis of more complex situations in which the eddy transfer of momentum plays a leading part.

This has been done by Prandtl by the use of a model which is obviously suggested by the simplest Maxwell picture of the generation of viscosity in a gas (*vide* Chapter I, p. 3). An eddy, loosely defined as a volume of fluid which has acquired the mean motion of its surroundings, is supposed to break away from the mean flow and to move in the z direction. In doing so it is assumed to conserve its momentum, and if ultimately it is reabsorbed into the main motion at a level $z+l$, the effect of the mixing will be to produce at that level an oscillation of velocity u' equal to

$$\bar{u}(z+l)-\bar{u}(z)\simeq l d\bar{u}/dz \qquad . \quad . \quad . \quad 4.11$$

In general, the root-mean-square values of the component fluctuations in a turbulent fluid tend to equality so that the z-oscillation w' may also be assumed to be expressible in the form $l\, d\bar{u}/dz$. This suggests that we may write

$$\tau=-\rho\overline{u'w'}=-\rho l^2\left(\frac{d\bar{u}}{dz}\right)\left|\frac{d\bar{u}}{dz}\right| \qquad . \quad . \quad . \quad 4.12$$

as the expression of the eddy shearing stress in terms of the velocity gradient and an undefined length l, which is obviously akin to the mean free path of the kinetic theory of gases. This quantity is called by Prandtl the ***turbulent mixing length*** (*Mischungsweg*).

We may consider the concept from a slightly different angle, giving rise to a somewhat different definition of the mixing length. Suppose that $E=E(z)$ is any transferable entity (such as momentum, heat or matter) whose mean value is constant over any (x, y) plane and which is supposed conserved during the motion of the eddy in the z direction. A mass of fluid starts from the plane $z=z_1$ and carries with it the mean value $\bar{E}(z)$ to the plane $z=z_2$. The mean rate of transfer of $E(z)$ across unit area of a plane perpendicular to z is

$$q=-\overline{w'[E(z_2)-E(z_1)]}\simeq-\overline{w'(z_2-z_1)}d\bar{E}/dz \quad . \quad 4.13$$

where w' is the transverse (eddy) velocity. We now suppose that a mean length l' exists which characterizes the mixing process and which is such that

$$l'\sqrt{(\overline{w'^2})}=\overline{w'(z_2-z_1)}. \quad . \quad . \quad . \quad . \quad 4.14$$

The effect of the turbulence on the transfer of E is therefore represented by

$$q=-l'\sqrt{(\overline{w'^2})}\, d\bar{E}/dz. \quad . \quad . \quad . \quad . \quad 4.15$$

In particular, if we assume that momentum is conserved in the sense indicated above we may put $E=\rho\bar{u}$, and so equation 4.15 becomes

$$\tau=-\rho l'\sqrt{(\overline{w'^2})}d\bar{u}/dz \quad . \quad . \quad . \quad . \quad 4.16$$

which may be compared with equation 4.12 above.

Comparison with the expression $\mu\, d\bar{u}/dz$ for the shearing stress due to viscosity shows that a suitable expression for the eddy viscosity or interchange coefficient (*vide* p. 30) is

$$K(z)=A/\rho=\overline{w'l}=l^2\, d\bar{u}/dz \quad . \quad . \quad . \quad 4.17$$

—a formula due to Prandtl—or

$$K(z)=l'\sqrt{(\overline{w'^2})}$$

using the second definition of the mixing length. (It is assumed that constants of proportionality are absorbed in the l's.) The lengths l and l' defined above are not necessarily equal, nor does it follow that the mixing length is independent of the property being transferred, although it is evident on grounds of the general similarity of transfer phenomena in turbulent fluids that l must at least be of the same order of magnitude for all properties.

At this stage it is not immediately evident that anything significant has been achieved. Obviously the analysis has not aimed at or attained anything like the precision or the detail of the kinetic theory of gases. The means have been evaluated in the crudest possible manner and no account has been taken of the fact that the value of $\overline{u'w'}$ depends essentially upon the degree of correlation existing between u' and w'. Despite this, the analysis, by bringing into prominence the length l, has supplied a useful tool for the resolution of many complex problems, chiefly because the mixing length, for large Reynolds numbers, turns out to be virtually independent of bulk factors such as the magnitude of the mean motion and the viscosity of the fluid. The reader must be warned against taking the Prandtl or any other model literally—the sole purpose of such models is to suggest how the complicated features of turbulent flow may become amenable to calculation by a suitable preliminary dissection process, and the approach is by analogy rather than by an exact analysis of a mechanical model. The success of the concept really lies in the fact that only very simple assumptions regarding the functional form of l are required in order to obtain a workable method of analysis of many aspects of turbulent flow.

Momentum-Transfer and Vorticity-Transfer Hypotheses. An outstanding difficulty in the Prandtl method is the basic hypothesis that momentum is conserved as the eddy moves from one level to another. This is tantamount to assuming that the oscillations of pressure (which certainly exist in turbulent flow and are not negligible) do not effect the mean transport of momentum. This is not evident *a priori*.

Considerations of this kind led Taylor to put forward, first in 1915 and again in 1932, the theory that it is vorticity and not momentum which is conserved in the process, although the final result is a transfer of momentum. This is based on the fact that if the turbulent motion is two-dimensional in the plane (x, y), one component of vorticity is unaffected by local variations of pressure and is therefore a suitable conservative entity for the mixing process described above. The effect of the substitution of vorticity for momentum is best seen by considering the expression for the rate at which momentum is communicated to unit volume of the fluid by turbulence. On the momentum transport theory this is

$$\frac{\partial\tau}{\partial z}=\rho\frac{\partial}{\partial z}\left\{l^2\left(\frac{d\bar{u}}{dz}\right)\left|\frac{d\bar{u}}{dz}\right|\right\} \quad . \quad . \quad . \quad . \quad 4.18$$

or, introducing the interchange coefficient $K(z)=l^2\, d\bar{u}/dz$ from equation 4.17,

$$\frac{1}{\rho}\frac{\partial\tau}{\partial z}=\frac{\partial}{\partial z}\left\{K(z)\frac{d\bar{u}}{dz}\right\} \quad . \quad . \quad . \quad . \quad 4.181$$

We now compare this expression with that obtained on the vorticity transport theory.

The effect of the transport of vorticity on the distribution of momentum is seen by considering, as before, the case in which the mean flow is entirely along the x-axis and depends only on z, and the turbulence is assumed constant in the x-direction. Neglecting viscous terms in comparison with the eddy terms, the equation of motion is

$$\frac{\partial u}{\partial t}=-\frac{1}{\rho}\frac{\partial p}{\partial x}-\frac{\partial}{\partial x}(\tfrac{1}{2}u^2+\tfrac{1}{2}w^2)-w\eta$$

where η is the vorticity $\partial u/\partial z-\partial w/\partial x$. Taking means, we have

$$\frac{\partial\bar{u}}{\partial t}=-\frac{1}{\rho}\frac{\partial\bar{p}}{\partial x}-\overline{w'\eta'}$$

where η' is the turbulent vorticity. This equation shows that the mean velocity at any level is increasing or decreasing at a rate $-\overline{w'\eta'}$ or that momentum is being increased or decreased at the rate $-\rho\overline{w'\eta'}$ per unit volume because of the turbulence.

The term $\overline{w'\eta'}$ may be expressed in terms of the mixing length; since η is a conservative transferable property we have, putting $E=\eta$ in equation 4.15

$$\rho\overline{w'\eta'}=\rho l'\sqrt{(\overline{w'^2})}d\bar{\eta}/dz=\rho l'\sqrt{(\overline{w'^2})}d^2\bar{u}/dz^2 \qquad 4.19$$

since $\bar{\eta}=d\bar{u}/dz$ by definition. Equation 4.19 may be compared with equation 4.18, or writing 4.19 in the interchange coefficient form by taking $K(z)=l'\sqrt{(\overline{w'^2})}$, we have

$$\frac{1}{\rho}\frac{\partial\tau}{\partial z}=K(z)\frac{d^2\bar{u}}{dz^2} \quad . \quad . \quad . \quad . \quad . \quad 4.191$$

as the counterpart of equation 4.181. There is thus a fundamental difference between the two theories for the rate at which momentum is communicated to unit volume of the fluid, for equations 4.181 and 4.191 are identical if and only if K=constant. As we have seen, this case is not applicable to most problems arising in the lower atmosphere.

Space does not permit further discussion here of the relative merits of the two theories and the reader is referred to Goldstein's *Modern Developments in Fluid Mechanics* (Oxford, 1938) for an account of these matters. This is one of the many cases in which meteorology must await the final verdict of the fluid motion laboratories, for observations in the lower atmosphere are hardly suitable for the resolution of a delicate and intricate question of this type. The development of the vorticity-transport theory leads to equations of considerable complexity and there is little doubt that Prandtl's method, however questionable its foundations may be, is much the easier to apply to the problems of the lower atmosphere. We shall therefore confine attention here to the application of Prandtl's concept to meteorology.

2. THE MIXING LENGTH IN RELATION TO THE VELOCITY PROFILE

We may examine the validity of Prandtl's approach to the problem in two ways: (i) we can evaluate l from observations with the aid of one of the equations 4.11, 4.12 or 4.17 and see if the remaining expressions are consistent with this result, or (ii) we may make some simple hypothesis regarding l (e.g. $l \propto z$) and test if such an assumption leads to results agreeing with observation. Both methods, and particularly the second, have been employed.

A fundamental relation for l may, however, be obtained immediately. It is reasonable to assume that l must depend, in the first instance, chiefly on the shape of the velocity profile in the immediate neighbourhood of the reference point. Since the mean velocity can be eliminated by taking axes moving with the fluid without affecting the eddy transfer of momentum, this means that l must be a function of $d\bar{u}/dz$, $d^2\bar{u}/dz^2$, . . . , and disregarding derivatives higher than the second we have on dimensional grounds

$$l = k\frac{d\bar{u}}{dz} \Big/ \frac{d^2\bar{u}}{dz^2},$$

where k is a pure number known as *Kármán's constant*. This result was first derived by Kármán on the assumption of dynamical similarity between the factors which determine the stresses at two points in a fluid. The numerical value of k has been determined by Nikuradse (1932) and others to be about 0·4 for flow on the laboratory scale.

Boundary Conditions and the Effects of Surface Roughness. At this stage it is necessary to discuss a problem of considerable uncertainty, namely, the exact form of the lower boundary condition. Not only is the surface of the earth normally covered with obstacles of varying height (such as grass, trees, houses and hills), but in addition it is often a matter of difficulty to determine the height at which the wind speed is effectively zero (e.g. in a field of corn). Problems of this type, however, have been extensively studied in controlled conditions in aerodynamical laboratories with results which have important applications to meteorology.

Aerodynamically Smooth and Rough Surfaces. In turbulent flow near a rigid boundary which is 'smooth' in the ordinary sense of the word (such as a polished surface) it is possible to distinguish three regions of flow:

(i) the laminar sub-layer immediately adjacent to the wall, in which the velocity gradient is very large and viscous stresses predominate;

(ii) an overlying boundary layer in which the viscous stresses and the Reynolds stresses are of equal magnitude;

(iii) the main turbulent mass of the fluid in which the Reynolds stresses predominate.

According to the fundamental investigations of Nikuradse (1933) and Schlichting (1936) on flow in smooth and artificially roughened pipes, the effect of surface roughness on resistance depends primarily on the ratio of the size of the irregularities to the depth of the laminar sub-layer. At low Reynolds numbers or for small irregularities the roughness elements, being completely submerged in this layer, make no effective contribution to the total friction, and in these circumstances the surface is said to be *aerodynamically smooth*. When the irregularities are of size comparable with the thickness of the laminar sub-layer an increase in resistance can be detected (*transitional flow*), and finally, with large roughness elements or large Reynolds numbers the resistance rises considerably and the surface is said to be *fully rough*. Viscosity plays a large part in determining the skin friction of an aerodynamically smooth surface but hardly enters at all in the case of a fully rough surface, for in these circumstances the irregularities behave like bluff bodies whose resistance is almost entirely form drag, i.e. determined mainly by the difference in pressure between the nose and the tail of the obstacle. Thus whether a surface is smooth or rough depends on both the size of the roughness elements and the magnitude of the mean velocity.

It has been shown in Chapter III, p. 41, that we may consider, without serious error, the eddy shearing stress τ to be independent of height near the ground and thus equal to its value at the surface (τ_0). It is convenient here to introduce the so-called friction velocity (*Schubspannungsgeschwindigkeit*) v_* defined as

$$v_* = \surd(\tau_0/\rho).$$

Since in Prandtl's theory $\tau = \rho l^2 (d\bar{u}/dz)^2$ it follows that v_* is of the form $l\, d\bar{u}/dz$ (or, more accurately, $v_* = \surd(|\overline{u'w'}|)$), so that the friction velocity is a quantity similar to the eddy velocities (and so approximately proportional to the mean velocity) and of the same order of magnitude. Obviously, in this theory the assumption that τ is independent of height necessarily implies also the constancy of v_* and also of the eddy velocities in the shallow layer concerned.

It is a matter of observation that the apparent shear stress in a turbulent fluid is approximately proportional to the square of the velocity and the friction velocity is the velocity for which this relation is exact. It is thus a more natural reference velocity for the layer in which τ is constant than, say, the mean velocity at an arbitrary fixed height and the Reynolds number appropriate to flow in this region is thus v_*z/ν. For a surface artificially roughened with grains of sand of average diameter ε Nikuradse has given the following critera for the three types of flow distinguished above:

aerodynamically smooth flow	$v_*\varepsilon/\nu<4$
transitional flow	$4\leqslant v_*\varepsilon/\nu\leqslant 75$
fully rough flow	$v_*\varepsilon/\nu>75$

The Logarithmic Profiles for Smooth and Rough Flow. Apart from the sterile assumption of a constant mixing length, the simplest hypothesis which we may make regarding l is to put $l\propto z$, and Nikuradse has shown that for pipe flow the straight line $l=kz$, where k is Kármán's constant ($c.$ 0·4), is the tangent to the $l=l(z)$ curve at the origin. We have then, near the boundary, the differential equation of the profile,

$$\frac{1}{v_*}\frac{d\bar{u}}{dz}=\frac{1}{kz} \quad . \quad . \quad . \quad . \quad . \quad . \quad 4.21$$

from the definition of v_* and the invariability of τ with height in the shallow layer considered. The boundary condition on $z=0$ requires careful consideration; for the large Reynolds numbers concerned the thickness of the laminar sub-layer is negligible, but we cannot neglect the fact that $d\bar{u}/dz$ attains very large values in this region. We therefore drop the too-restrictive 'zero slip' condition (i.e. $\bar{u}=0$ on $z=0$) and substitute instead the condition $\lim_{z\to 0} d\bar{u}/dz=\infty$. The solution of equation 4.21 may then be arranged in the form

$$\bar{u}/v_*=(1/k)\log_e(v_*z/\nu)+\text{const.} \quad . \quad . \quad . \quad 4.22$$

and evaluating the constant from observations in smooth pipes (Nikuradse, 1932) we have, for $k=0·4$

$$\bar{u}/v_*=2·5\log_e(v_*z/\nu)+5·5=2·5\log_e(9v_*z/\nu)\text{ approx.} \quad 4.221$$

a formula known as the *universal logarithmic velocity profile for aerodynamically smooth flow*. (Somewhat surprisingly, considering the restrictive nature of the assumption τ=constant, this expression agrees closely with observations not only near the wall but right up to the centre of the pipe.)

In the case of fully rough flow viscosity plays a negligible part in the determination of the skin friction, so that the only two lengths which enter into the argument are z (or l) and ε. The solution of equation 4.21 in this case is of the form

$$\bar{u}/v_{*}=(1/k)\log_e z/\varepsilon+\text{const.} \quad . \quad . \quad . \quad 4.23$$

This is usually written as

$$\bar{u}/v_{*}=(1/k)\log_e (z/z_0) \quad . \quad . \quad . \quad 4.231$$

where z_0 is a function of the size of the roughness elements (ε) called the *roughness length*. Equation 4.23 is known as the *universal logarithmic velocity profile for fully rough flow*. Its validity has been demonstrated by Schlichting (1936) in a series of laboratory trials.

Equations 4.22 and 4.23 may be regarded as providing criteria which determine whether a surface is aerodynamically smooth or rough. This is particularly appropriate to meteorological problems where it is clearly impossible to speak with any confidence of the 'average size' of the roughness elements. For artificial surfaces roughened with grains of sand it has been shown that z_0 is about one-thirtieth of the average diameter of the grains. Schlichting (loc. cit.) widened the scope of the investigation by examining the details of flow over flat plates roughened with elements such as spheres, rivet heads, cones, &c., of different sizes, aspects and spacings, and in this way was able to define an 'equivalent sand roughness' and to show that equation 4.23 holds for a wide range of boundary conditions.

The General Logarithmic Profile for Smooth and Rough Flow. A difficulty in the use of equation 4.231 is that it implies a discontinuity in the profile shape as the roughness parameter approaches zero, i.e. equation 4.221 is not the limiting form of equation 4.231 as $z_0 \rightarrow 0$. It is also clearly desirable to formulate an expression for the velocity profile in which z_0 can take unrestricted values, including zero. Expression 4.231 for the rough flow profile may be written in the form

$$\bar{u}/v_* = (1/k) \log_e (v_* z/N) \quad . \quad . \quad . \quad 4.232$$

where $N = v_* z_0$, an expression which closely resembles that for smooth flow with N replacing ν. As will be seen from Table I, Appendix, for winds blowing over average terrain, v_* is usually about one-tenth of the mean velocity and z_0 is generally a few centimetres, so that N is of the order of 10^2 cm.^{2}sec.$^{-1}$ or greater, and is thus very much greater than ν. We call the constant N the *macro-viscosity* of the flow to distinguish it from the eddy viscosity proper.

The above considerations suggest that a generalized profile for both smooth and rough flow may be defined as

$$\bar{u}/v_* = (1/k) \log_e \{v_* z/(N + \tfrac{1}{9}\nu)\} \quad . \quad . \quad 4.233$$

This expression satisfies all the requirements, for as $z_0 \rightarrow 0$, equation 4.233 tends to the smooth-surface profile (4.221), and if $N \gg \nu$, equation 4.233 is indistinguishable from the ordinary rough-surface profile (4.231). In this expression z_0 may be allowed to take unrestricted values, including zero.

The macro-viscosity is obviously akin to the kinematic viscosity. In the kinetic theory of gases, ν=const. cl, where c is the mean molecular velocity and l is the mean free path, which may be regarded as specifying the average geometrical configuration of the molecular field (*vide* Chapter I). The macro-viscosity is similarly related to u', the eddy velocity, and to z_0, which in this case may be regarded as specifying the geometrical configuration of the surface and therefore, in some way, the distribution of the eddies in the flow near the surface.

Applications to the lower atmosphere. The logarithmic profile for aerodynamically smooth surfaces (eq. 4.221) has been compared with observations in the atmosphere by Rossby (1936), who concludes that in moderately light winds the sea surface is aerodynamically smooth. The same test has also been applied by P. A. Sheppard (unpublished) to the profiles obtained by Best (op. cit.) for heights between 2·5 cm. and 5 m. and wind velocities between 1·5 and 4 m./sec. in adiabatic gradient conditions over the short-cropped grass of a cricket field. From the fact that the velocity gradients observed by Best are much greater than those appropriate to an aerodynamically smooth surface, Sheppard concludes that a closely cut lawn is 'rough' at these relatively low wind speeds. *A fortiori*, all natural surfaces, such as downland or cultivated areas, must be rough at normal wind speeds, at least in adiabatic gradient and lapse conditions. Whether the same conclusion is valid for inversions is not known.

We shall therefore confine attention to equation 4.23, i.e. to the profile for fully rough flow. A difficulty immediately arises regarding the origin of z. In his aerodynamic investigations Schlichting defined this as an imaginary surface which could replace the actual rough wall without altering the fluid volume. For the atmospheric case it is customary to introduce a *zero plane displacement* (d) to provide a datum level above which active turbulent interchange commences and the logarithmic profile might be expected to hold. The modified form of equation 4.23 is then

$$\bar{u}/v_{*}=(1/k)\log_e\{(z-d)/z_0\} \quad . \quad . \quad . \quad 4.24$$

where d is of the order of magnitude of the depth of the still air trapped among the roughness elements of the surface and $z \geqslant z_0+d$. This equation is not deducible from the original differential equation 4.21 if z_0 and d are regarded as independent arbitrary constants, for equation 4.21 is of the first order and therefore its solution can contain at the most only one arbitrary constant. At this stage, equation 4.24 must be looked upon as an empirical modification of the logarithmic profile proper.

Determination of z_0 and d. Equation 4.24 has been used by various workers in this subject. Sverdrup, following Rossby and Montgomery (1935, 1936), assumes $d=-z_0$, thus employing only one arbitrary constant. Paeschke (1937), in an analysis of profiles over various natural surfaces (snow, grassland, cultivated land), takes d to be a positive quantity equal to the average of the measured heights of the roughness elements. More recently E. L. Deacon (1949) working at Porton concludes that in conditions of neutral stability the logarithmic law can represent the profile between heights of 1 m. and 13 m. over grass of various length with great accuracy, provided that both z_0 and d are chosen independently to give the best fit.

A summary of some of these results is given in the Appendix. For exceptionally smooth surfaces, such as mud flats in tidal estuaries or snow or ice in a very level condition, values of z_0 of the order of 10^{-3} to 10^{-2} cm. appear to be appropriate. Short grass usually requires values of z_0 less than 1 cm., while long grass necessitates taking z_0 to lie between 1 cm. and 10 cm., and in this case a zero-plane displacement is also necessary. In general, most writers agree as regards the order of magnitude of z_0, but as might be expected, there is a considerable spread in the actual values obtained.

Determination of l and k near the Ground. A detailed analysis of the form of the mixing length near the ground has recently been published by P. A. Sheppard (1947), using data obtained at Porton from measurements of the eddy shearing stress and of the aerodynamic drag of the earth's surface. The site chosen was exceptionally smooth, being an isolated area of concrete about 160 m. in diameter on Salisbury Plain, and the value of the drag was deduced from observations on the deflection of a small horizontal test surface floating in a bath of oil. In his reduction of the results, Sheppard used equation 4.24 in the form

$$\bar{u}/v_*=(1/k)\log_e\{(z+z_0)/z_0\} \quad . \quad . \quad . \quad 4.241$$

which implies that $u=0$ on $z=0$ and also that

$$l=k(z+z_0). \quad . \quad . \quad . \quad . \quad . \quad . \quad 4.25$$

Since for any given terrain and temperature gradient z_0 is a fixed length of the order of centimetres, whereas z may take any values up to several tens of metres, this implies that the influence of z_0 on the mixing length is negligible for heights greater than a few metres.

Sheppard's results are of considerable interest in the

study of turbulence near the ground for they constitute the most direct measurements yet made of both l and k in the atmosphere. The conclusions reached may be summarized as follows:

(i) in conditions of small lapse rate the straight line

$$l=0\cdot45z$$

(corresponding to $k=0\cdot45$) is a good fit to the observations made over the concrete surface for heights up to about 1 m. Above that height the mixing length increases more rapidly than the height.

(ii) Over the range $0 \leqslant z \leqslant 2$ m., and for the same conditions as in (i) above, the variation of mixing length with height is well represented by

$$l=0\cdot25z^{1\cdot15}$$

(iii) There is evidence to show that in lapse conditions the mixing length increases more rapidly than the height, while in inversions the increase is less rapid than the height.

(iv) The Kármán constant k and the roughness length z_0 are both functions of stability over a short cropped grass surface.

The data obtained are:

Temperature gradient	z_0 (cm.)	k
(Superadiabatic) lapse	0·006	0·61
Adiabatic gradient	0·28	0·40
Inversion	0·80	0·22

(A variation of z_0 with temperature gradient in the above sense had also been deduced by Sutton in 1936.)

In assessing the value of these conclusions it should be noted that two important assumptions have been made: (i) that the deflection of the test surface is a reliable measure of the drag of the ground itself, and (ii) that the logarithmic velocity profile is equally valid for both adiabatic and non-adiabatic gradients. The first assumption is the reason for the experimental technique used and it is not easy to see what other means can be adopted for a direct measurement of the drag, but the second assumption is not as necessary and may account, in part, for the variations observed in l and k as the temperature gradient changes. This is of particular importance since it is now known that the logarithmic law holds for neutral equilibrium and power laws hold only in other conditions.*

However, the main result of the above investigations is that in spite of the enormous difference in scale, atmospheric and wind tunnel data on the transfer of momentum are in reasonable accord. This is important for future work, for the analysis of turbulent flow is much easier under controlled conditions in the laboratory than in the open, and the application of aerodynamical theories to meteorology may thus be expected to be fruitful of results.

The absence of the viscosity from the expressions for the profile over fully rough surfaces should not be taken as indicating that molecular forces play no part in atmospheric turbulence. Ultimately, all transfer of momentum must be by viscosity and what the above analysis indicates is that the appropriate boundary condition for winds in the lower atmosphere is one in which the direct effect of viscosity may, in the first instance, be neglected, but the spread of the influence of the surface upwards must, in the final analysis, involve viscosity. This is likely to be particularly true when the flow borders on the non-turbulent state, i.e. in conditions of large inversions.

3. STATISTICAL THEORIES OF TURBULENCE

The theory discussed above is based, however indirectly, upon a mechanical picture of turbulent flow which conflicts with intuition at many points. In reality, mixing in a fluid cannot be the discontinuous process which the Prandtl theory postulates since the velocities and the motions of the particles in turbulent flow are continuous. In particular it is impossible to accept literally the idea that diffusion proceeds by a series of definite jumps, each of which is unrelated to those which have preceded it.

* Sheppard's results have recently been discussed by R. B. Montgomery; see *Q.J. Roy. Meteor. Soc.*, 73 (1947), pp. 459–62.

These objections do not apply to the theory of *diffusion by continuous movements*, introduced by Taylor in 1922, in which the motion of a particle at any instant is partly determined by its previous history. The conception of an event occurring in a definite time series and therefore being related to those which follow it or precede it in the same system has led to the introduction of *auto-correlation coefficients* in modern statistical work. Space does not permit a discussion of this aspect of the matter here and for further information the reader is referred to Kendall's *Advanced Theory of Statistics* (Vol. II), from which we borrow an illustration which applies particularly well to the theory of turbulence. Suppose a well-sprung car is moving over a rough road. On striking a bump the car will oscillate with a damped vibration which would ultimately die out were it not reinforced by the car meeting another bump, and so on. The motion of the car is thus determined by (i) its average speed, (ii) the characteristics of its spring and vibration damper system, and (iii) the size and distribution of the bumps which have been and which are being encountered. If the bumps are of roughly equal size and evenly distributed, the oscillatory motion of the car will have a recognizable regularity, e.g. the root-mean-square value of the amplitudes of the oscillations might be approximately constant over a long period. This regularity of the motion (in the statistical sense) is mainly a consequence of the internal structure of the vehicle, but the existence of the oscillations depends upon the random external impulses supplied by the bumps in the road.

A similar state of affairs holds in the case of the wind blowing over a natural surface. A disturbance due to a tree or similar obstacle effects the motion of the particles of air in the immediate neighbourhood, but such effects are transitory and the additional motion is ultimately dissipated by viscosity. The flow, as a whole, is obviously made up of a succession of such disturbances imposed on a mean motion, and while the statistical regularity of the motion is ultimately to be attributed to the internal structure of the fluid (i.e. to its viscosity and density), the existence of such disturbances depends upon a succession of random events.

Taylor's Theorem. We now prove a remarkable theorem due to Taylor (1922) relating to diffusion in a statistically uniform field of flow, i.e. one in which the mean eddying energy, $\rho\overline{u'^2}$, is constant with respect to time and space. We proceed to follow the history of a group of particles in this field. Let R_ξ be the correlation between the eddy velocities u'_t at time t and $u'_{t+\xi}$ at time $t+\xi$ which affect these particles. Then by definition of a correlation coefficient,

$$R_\xi=\overline{u'_t u'_{t+\xi}}/\overline{u'^2} \quad . \quad . \quad . \quad . \quad . \quad 4{\cdot}31$$

Integrating with respect to ξ,

$$\overline{u'^2}\int_0^t R_\xi d\xi=\int_0^t \overline{u'_t u'_{t+\xi}}d\xi=\overline{u'_t\int_0^t u'_{t+\xi}d\xi}=\overline{u'_tX} \quad . \quad 4{\cdot}32$$

where X is the distance travelled by the particle as a consequence of the eddy motion during a time interval t.

Thus

$$\overline{u'^2}\int_0^t R_\xi d\xi=\overline{u'_tX}=\tfrac{1}{2}\frac{d}{dt}\overline{X^2} \quad . \quad . \quad . \quad . \quad 4{\cdot}33$$

or

$$\overline{X^2}=2\overline{u'^2}\int_0^T\int_0^t R_\xi d\xi\ dt \quad . \quad . \quad . \quad . \quad . \quad 4{\cdot}34$$

Thus in considering the diffusion of particles originally concentrated on the (x, z) plane at time $t=0$, it has been shown that the standard deviation $\overline{X^2}$ of the distance travelled in time T can be expressed entirely in terms of T, $\overline{u'^2}$ and R_ξ. The analysis also yields a length l, defined by

$$\overline{Xu'}=l_1\sqrt{(\overline{u'^2})}=\overline{u'^2}\int_0^{t_0}R_\xi d\xi$$

i.e.

$$l_1=\sqrt{(\overline{u'^2})}\int_0^{t_0}R_\xi d\xi \quad . \quad . \quad . \quad 4{\cdot}35$$

where t_0 is so great that R_ξ is zero or negligibly small for $\xi>t_0$. This length clearly is analogous to the mixing length defined in the first part of this chapter, but is not dependent upon mixture, and the theory of diffusion given above is equally valid if mixture does not take place.

If T is small, so that effectively $R_\xi=1$, equation 4.34 yields

$$\sqrt{(\overline{X^2})}=\sqrt{(\overline{u'^2})}.T$$

i.e. the deviation is initially proportional to the time. In a turbulent fluid we anticipate that $R_\xi \rightarrow 0$ as $\xi \rightarrow \infty$, and if we suppose in addition that $\lim_{t\rightarrow\infty} \int_0^t R_\xi d\xi$ is finite and equal to I we have, for sufficiently great T,

$$\overline{X^2}=2\overline{u'^2}IT \quad . \quad . \quad . \quad . \quad . \quad 4.36$$

where I, from 4.35, is of the form $l_1/\sqrt{(\overline{u'^2})}$ and is constant. Hence we may write equation 4.36 in the form

$$\sqrt{(\overline{X^2})}=\sqrt{(2KT)} \quad . \quad . \quad . \quad . \quad 4.37$$

where K is of the form $l_1\sqrt{(\overline{u'^2})}$ and is therefore an interchange coefficient (cf. eq. 4.191, p. 49). Equation 4.36 thus indicates that in a fluid of constant diffusivity the standard deviation of the distances travelled by the particles is proportional to the square root of the time which has elapsed since they were originally concentrated on the plane (x, z). This law was obtained by Einstein in his discussion of Brownian motion and has been verified by Perrin, but since in the atmosphere K cannot be regarded as constant, it has no particular significance in meteorology.

In later papers (1935, 1936) Taylor considerably extended this theory and his work has been followed by Kármán and Howarth (1938), Robertson (1940) and Synge and Lin (1943), while Sutton (1932, 1934) has applied Taylor's theorem to diffusion in the lower atmosphere. The meteorological applications will be dealt with later (Chapter V) and we shall here conclude with a brief sketch of the main lines of development of the pure theory.

Isotropic Turbulence and the Dissipation of Energy. Taylor, in his later papers, considerably simplified the problem by the introduction of the concept of *isotropic turbulence*, i.e.

flow in which the average value of any function of the velocity components and their spatial derivatives is unaltered if the axes of reference are rotated or are reflected. It is also convenient to introduce a new correlation coefficient, R_y, defined as the correlation between simultaneous values of u' distant y apart, and a length l_2 defined by

$$l_2=\int_0^\infty R_y\,dy.$$

This length l_2 differs from the length l_1 defined before in the way in which Eulerian hydrodynamics differs from the Lagrangian system, i.e. l_1 refers to the group of particles whose history is being traced, whereas l_2 refers to the whole field of flow at any instant. We may speak of l_2 as the *scale of the turbulence*, since it is a measure of the average size of the eddies which make up the pattern of flow.

The analysis is particularly appropriate to the case in which the turbulence is decaying, e.g. additional turbulence might be introduced into a wind tunnel by the introduction of a regularly spaced grid of parallel bars near the inlet end of the working section and the properties of the flow studied at various points downstream. The general expression for the mean rate of dissipation of energy in isotropic turbulent flow is*

$$\bar{W}=6\mu\left\{\overline{\left(\frac{\partial u'}{\partial x}\right)^2}+\overline{\left(\frac{\partial v'}{\partial x}\cdot\frac{\partial u'}{\partial y}\right)}+\overline{\left(\frac{\partial u'}{\partial y}\right)^2}\right\}$$

which in virtue of the special isotropic relations assumed reduces to

$$\bar{W}=7\cdot5\mu\overline{(\partial u'/\partial y)^2}$$

Now it may be shown that

$$R_y=\frac{\overline{u'u'_y}}{\overline{u'^2}}=1-\frac{1}{2!}\frac{y^2}{\overline{u'^2}}\overline{\left(\frac{\partial u'}{\partial y}\right)^2}+\ldots$$

* For the details of the analysis, see Taylor's original papers, or Goldstein, *Modern Developments in Fluid Dynamics*, I, pp. 221, &c.

so that the curvature of the R_y curve near the origin is a measure of $\overline{(\partial u'/\partial y)^2}$ and therefore

$$\overline{\left(\frac{\partial u'}{\partial y}\right)^2}=2\overline{u'^2}\lim_{y\to 0}\left(\frac{1-R_y}{y^2}\right).$$

Now $(1-R_y)/y^2$ is the reciprocal of the square of a length; we call this length λ and we have

$$\bar{W}=15\mu\,\overline{u'^2}/\lambda^2 \quad . \quad . \quad . \quad . \quad . \quad . \qquad 4{\cdot}38$$

The length λ is usually termed the *microscale of turbulence* and may be regarded as indicating the average size of the smallest eddies which are responsible for the greater part of the dissipation. It may also be shown that λ is related to the eddy velocity u' and l_2, the scale of the turbulence, by the equation

$$\lambda^2=\text{constant}\{l_2\nu/\sqrt{(\overline{u'^2})}\} \quad . \quad . \quad . \qquad 4{\cdot}39$$

so that ultimately

$$\bar{W}=\text{constant}\{\rho(\sqrt{\overline{u'^2}})^3/l_2\} \quad . \quad . \quad . \qquad 4{\cdot}310$$

These and other relations obtained by Taylor have been substantially verified by Simmons and others for the decay of isotropic turbulence in the wake of a grid.

Other Developments. Kármán and Howarth have introduced the concept of the correlation tensor defined as follows. For isotropic turbulence the correlations between the velocity components u_i at P and v_j at a second point P' become $\overline{u_i v_j}/\overline{u^2}$ which may be considered to be the components of a second-rank tensor R_{ij} having spherical symmetry and depending only upon r, the distance between P and P', and the time. It can be shown that R_{ij} can be expressed in terms of one scalar function only, namely the correlation coefficient between the velocity components at two points directed along the line joining the points.

Very little has been deduced theoretically concerning the functional form of the correlation coefficients. By making certain assumptions, Kármán and Howarth have shown that their fundamental correlation coefficient can be expressed as $\exp(-ar^2/\nu t)$, where a is a constant.

The most striking advance in the general theory in recent years undoubtedly is that initiated by A. N. Kolmogoroff and developed by G. K. Batchelor (1947, 1950). The impetus for a more rational approach than the crude empiricism of the mixing-length theories caused a number of mathematicians, including Heisenberg, Onsager and Weiszächer to put forward similar ideas about the same time.

Kolmogoroff's analysis embodies the fundamental concept that the eddying motions are characterized by a wide range of length scales. Energy continually passes from the large to the smaller eddies, ultimately to be absorbed into the random molecular motion by viscosity. In this process the direct influence of the larger eddies is felt less and less, so that there is a range of small eddies with properties common to all types of turbulence. Such eddies must be isotropic.

The energy exchange process is defined by two hypotheses, both applying to motions with large Reynolds numbers:

1. *The small-scale components of the motion depend only on the viscosity* (ν) *and the mean rate of dissipation of energy per unit mass of fluid* (ε).
2. *There is a sub-range of small eddies in which mean properties depend only on the mean rate of dissipation of energy per unit mass of fluid* (ε).

Dimensional analysis shows that the 'size' of the eddies directly influenced by viscosity is of the order of $(\nu^3/\varepsilon)^{\frac{1}{4}}$. In the lower layers of the atmosphere ε is about 5 cm.^{2}sec.$^{-3}$, so that the viscosity-influenced eddies must be very small, of the order of a centimetre at the most. The largest atmospheric eddies, on the other hand, are associated with length scales of the order of hundreds of metres, at least. Thus the subrange of small eddies to which Kolmogoroff's second hypothesis applies is very large in the atmosphere, which promises well for future applications. This is particularly so in problems of diffusion (see Chapter V).

So far, there have been few detailed applications of these ideas to meteorological problems, and most of the tests of

Kolmogoroff's theory have been by the way of wind-tunnel experiments. In Chapter V we shall consider mainly the application of Taylor's theorem to the problem of diffusion near the ground, with the object of establishing a technique for resolving practical problems, such as that of estimating the rate of evaporation from natural surfaces.

REFERENCES

For more detailed accounts of the theories described in this chapter the reader should consult

Aerodynamic Theory, ed. W. F. Durand, Vol. III, Section G, 1934.
Modern Developments in Fluid Dynamics, ed. S. Goldstein, Vol. I (1938),
in both of which full references are given to the original papers.

Batchelor, G. K. 1947. *Proc. Cambridge Phil. Soc.*, 43, 533
1950. *Q.J. Roy. Meteor. Soc.*, 76, 133
Sutton, O. G. *Micrometeorology*. New York (1953)

CHAPTER V

DIFFUSION IN THE ATMOSPHERE NEAR THE GROUND

AS indicated in Chapter I, the turbulent diffusion of suspended matter in the atmosphere has profound implications for our daily lives; in fact, it may be said that life as we know it could hardly proceed if such large-scale mixing did not take place. From the purely economic aspect, the study of atmospheric diffusion is of importance in questions relating to atmospheric pollution, agricultural meteorology and hydrology, while in military matters its significance in chemical warfare and in the problem of the screening of targets by smoke hardly needs emphasizing. More recently, the subject has been recognized as playing a major part in problems of the propagation of high-frequency electromagnetic waves over the earth. A considerable amount of the information available on diffusion emanates from the Meteorology Section of the Chemical Defence Experimental Station, Porton (Sutton, 1947), and the present chapter is mainly concerned with the theoretical aspects of the work carried out there from 1921 onwards.

1. STATEMENT OF THE PROBLEM AND THE EXPERIMENTAL DATA

The main technical problem of atmospheric diffusion may be stated thus: Given a source which emits matter in the form of a gas or a fine suspension (i.e. as a cloud of particles whose terminal velocity is negligible compared with the speed of the wind), to devise means whereby the rate of diffusion, and hence the properties of the cloud, can be predicted from a prior knowledge of the state of the atmosphere itself. The source of matter might be a factory chimney or a military smoke generator, a gas shell or bomb, or a pool of liquid yielding vapour by evaporation. The form of the resulting cloud will depend not only on the nature and magnitude of turbulent diffusion but also on the type and geometrical configuration of the source; thus gas shells and bombs are effectively instantaneous sources and factory chimneys are continuous sources and all may occur singly or combined to form point, line or area sources.

Instantaneous and Time-Mean Measures of Concentration. It has already been stated (Chapter II) that air flow near the surface of the earth includes oscillations having periods varying from a fraction of a second to many minutes. This has an important bearing on measurements of concentration (i.e. of the mass of suspended matter per unit volume) taken downwind of a continuous source. At short distances (up to 1,000 m. at least) from such a source there is a marked difference between instantaneous and time-mean samples of the concentration. At any instant the cloud from a continuous point source in a steady wind is shaped like a long narrow cone and a graph of instantaneous values of concentration against distance from the centre of the cloud at a fixed distance downwind and a fixed height above the surface is of the 'cocked-hat' type with a relatively high peak and a narrow base. If, however, continuous samples are taken over several minutes at the same positions the crosswind concentration curve, while retaining the same general shape, is considerably broader and has a lower peak value. This effect is caused by a slow swinging of the narrow 'instantaneous' cone over a wider front, so that the 'instantaneous aspect' of the cloud may be attributed mainly to the action of the smaller (short period) eddies while the 'time-mean aspect' is the resultant of the superposition of the larger (long period) eddies on the small-scale turbulence. The effect is considerable for lateral diffusion near the ground (at 100 m. from the source the 'instantaneous' width of the cloud from a continuous point source in adiabatic gradient conditions is about 20 m., while the 'time-mean' width is about 35 m.), but does not seem to be so pronounced for diffusion in the vertical. The data given below are time-mean values based on sampling periods of at least 3 minutes, since observations showed that although the front covered by the cloud increases rapidly as the period of sampling increases up to 2 minutes, the change thereafter is very small and consistent results are obtained for any period of sampling lasting not less than 3 minutes in steady conditions.

Definitions. The following technical terms are used in studies of atmospheric diffusion:

Strength of Source. This is the rate of emission of gas or particulate matter, usually expressed in g./sec. for a continuous point source and g./sec. metre for a cross-wind line source.

Width and Height of Clouds. The *width* of a cloud from a continuous point source is the distance between points on the skirts of the crosswind concentration curve at which the concentration is a fixed fraction, normally one-tenth, of the peak value. Similarly, the *height* of the cloud is defined as the vertical distance from the ground to the point at which the concentration has fallen to one-tenth of the value on the ground.

Concentration. This is the density of suspended matter, usually expressed in mg./cu. m. for convenience.

Experimental Data for Adiabatic Gradient Conditions. The following data are the mean results of many trials with both smoke and gas clouds over level downland. No difference could be detected between the rates of diffusion of gases and smokes.

1. The concentration at any point in a continuously generated cloud is directly proportional to the strength of the source, provided that the source itself does not materially interfere with the natural air flow (e.g. by producing intense local convection currents).
2. For a given strength of source the mean concentration at any point in a continuously generated cloud is approximately inversely proportional to the mean wind speed measured at a fixed height.
3. The time-mean width of the cloud from a continuous point source, measured at ground level, is about

35 m. at 100 m. downwind of the source and shows only very small variations with the mean wind speed.

4. The time-mean height of the cloud from an infinite crosswind continuous line source is about 10 m. at 100 m. downwind of the source and shows only very small variations with wind speed.
5. The central (peak) mean concentration from a continuous point source decreases with distance downwind x according to the law

$$\text{concentration} \propto x^{-1\cdot76}$$

6. The peak (i.e. ground level) mean concentration from an infinite crosswind continuous line source decreases with distance downwind according to the law

$$\text{concentration} \propto x^{-0\cdot9}$$

7. The absolute values of the peak mean concentrations at 100 m. downwind are as follows:

Type of source	Strength	Mean wind at 2 m. height	Peak concentration
Continuous point .	1 g./sec.	5 m./sec.	2 mg./cu. m.
Continuous infinite line (across wind)	1 g./sec. metre	5 m./sec.	35 mg./cu. m.*

* This value differs slightly from that given by Sutton (loc. cit.) and is the result of a redetermination by K. L. Calder, giving different weights to certain experimental data.

The above data constitute a standard set of values to which any theory of atmospheric diffusion must conform. It is unfortunate that as yet no corresponding set has been published for non-adiabatic temperature gradients, but it should be emphasized that the general unsteadiness and erratic behaviour of the light winds which are associated with both large lapse rates and large inversions make the experimental study of atmospheric diffusion in these conditions a matter of considerable difficulty.

2. THE ADAPTATION OF THE STATISTICAL THEORY OF TURBULENCE TO DIFFUSION OVER SMOOTH SURFACES

The mathematical theory which has been used in the preparation of range tables for Chemical Defence consists of an application of Taylor's correlation theory (Chapter IV) to turbulent motion near the ground (Sutton, 1932, 1934, 1947).

The co-ordinate system and notation employed are those given in Chapter II, p. 16, i.e. x is measured in the direction of the mean wind, y across wind and z vertically, so that the mean wind is represented by a single velocity $\bar{u}$ along the x-axis, and is supposed to be steady and to vary only in the vertical. The gustiness components g_x, g_y, g_z (*vide* Chapter II) are assumed to be constant (but unequal) in the relatively shallow layers under consideration.

It has already been explained that very little is known theoretically of the form of the correlation coefficient R_ξ, whose accurate measurement in the atmosphere is obviously difficult. Since R_ξ is non-dimensional it is possible, by using arguments of a type familiar in fluid motion theory, to deduce a probable form for this function.

For diffusion in the vertical, by definition

$$R_\xi = \overline{w'_t w'_{t+\xi}} / \overline{w'^2}$$

where $\overline{w'^2}$ is constant. For flow near an aerodynamically smooth surface the diffusive properties of the motion may be expected to depend principally upon the mean eddying energy $\rho\overline{w'^2}$ and the viscosity μ. This suggests that R_ξ will involve, in the first instance, only the constants ρ, μ, $\overline{w'^2}$ and the time ξ, and since the only non-dimensional combination which can be formed from these quantities is $\mu/\rho\overline{w'^2}\xi$ or $\nu/\overline{w'^2}\xi$, we write

$$R_\xi = f(\nu/\overline{w'^2}\xi)$$

The terminal conditions for R_ξ are: $R_0 = 1$ and $R_\xi \rightarrow 0$ as $\xi \rightarrow \infty$, which suggests that for large ξ, R_ξ may behave like $(\nu/\overline{w'^2}\xi)^n$ or like $1 - \exp(-\nu/\overline{w'^2}\xi)$, &c., where n is some undetermined pure number. For ease of manipulation a simple power is to be preferred to an exponential function and the form ultimately chosen for smooth flow is

$$R_\xi=\left(\frac{\nu}{\nu+\overline{w'^2}\xi}\right)^n \quad . \quad . \quad . \quad . \quad . \qquad 5.21$$

where n is an undetermined positive constant. This expression is the starting-point of the theory.

The hypothesis is now made that mixing in a turbulent fluid is defined by a loss of correlation between the initial and final eddy velocities which affect a particle, and the mixing length is defined as the distance over which an eddy moves during the time necessary for the correlation coefficient to become negligibly small. If t_0 be this time, we have for the vertical diffusion coefficient, since $\overline{w'^2 t_0}\gg\nu$,

$$K(z)=\overline{w'l}=\overline{w'^2}\int_0^{t_0} R_\xi d_\xi \simeq \frac{\nu^n}{1-n}(\overline{w'^2}t_0)^{1-n} \quad . \qquad 5.22$$

and using Kármán's expression for the mixing length, namely,

$$l=k\frac{d\bar{u}}{dz}\Big/\frac{d^2\bar{u}}{dz^2}$$

together with the Maxwellian law of distribution of eddy velocities (Chapter II) we have

$$K(z)=\frac{(\frac{1}{2}\pi k^2)^{1-n}}{1-n}\nu^n\left\{\left|\frac{d\bar{u}}{dz}\right|^3\left|\frac{d^2\bar{u}}{dz^2}\right|^{-2}\right\}^{1-n} \quad . \qquad 5.23$$

i.e. the eddy diffusivity has been expressed in terms of the velocity profile, Kármán's constant and the parameter n. From 5.21 and 5.22 it follows that n must lie between 0 and 1, since K is essentially positive.

The explicit expression for n in terms of the velocity profile is obtained by making use of Schmidt's conjugate power-law theorem (Chapter III), viz. that near the ground

$$\tau/\rho=K(z)\, d\bar{u}/dz=\text{const.} \quad . \quad . \quad . \qquad 5.24$$

whence it follows from equation 5.23 that the profile which satisfies the condition $\bar{u}=0$ on $z=0$, is

$$\bar{u}=\bar{u}_1(z/z_1)^{n/(2-n)} \quad . \quad . \quad . \quad . \quad 5.25$$

where $\bar{u}_1$ is the mean wind speed at the fixed reference height z_1. This equation allows n to be determined from measurements of the velocity profile and, clearly n will be small for lapse conditions and tend to unity in large inversions.

From equations 5.23 and 5.24 it follows that

$$K(z)=\left[\frac{(\frac{1}{2}\pi k^2)^{1-n}(2-n)n^{1-n}}{(1-n)(2-2n)^{2(1-n)}}\right]\nu^n\bar{u}_1{}^{1-n}z^{2(1-n)/(2-n)}z_1{}^{-n(1-n)/(2-n)}$$

$$=a(n)\nu^n\bar{u}_1{}^{1-n}z^{2(1-n)/(2-n)}z_1{}^{-n(1-n)/(2-n)}, \text{ say} \quad . \quad 5.26$$

—the final explicit expression for the eddy diffusivity in terms of the height, Kármán's constant and the velocity profile only. The function $a(n)$ is tabulated for a range of values of n in the Appendix.

Equations 5.24 and 5.25 provide a test of the theory by enabling the constant shearing stress at a smooth surface to be calculated and compared with observation. The velocity profile near a rigid smooth boundary is given, to a close approximation, by the 'one-seventh power law', i.e.

$$\bar{u}=\bar{u}_1(z/z_1)^{1/7}.$$

From equation 5.25 it follows that $n=\frac{1}{4}$ and hence, using equations 5.25 and 5.26, with $k=0\cdot4$,

$$\tau/\rho=K(z)\,d\bar{u}/dz=0\cdot020\nu^{1/4}\bar{u}_1{}^{7/4}z_1{}^{-1/4}.$$

The experimental result is*

$$\tau/\rho=0\cdot0225\nu^{1/4}\bar{u}_1{}^{7/4}z_1{}^{-1/4}$$

which indicates that the form adopted for R_ξ is close to the truth for smooth flow.

* See *The Physics of Solids and Fluids*, by Ewald, Poschl and Prandtl, p. 282.

3. SOLUTION OF THE STEADY STATE INFINITE LINE SOURCE PROBLEM AND EXTENSION TO THREE DIMENSIONS

When the emission is from a long line of constant continuous sources placed across the direction of the mean wind the problem becomes two-dimensional and requires the solution of the partial differential equation for the mean concentration $\bar{\chi}(x, z)$

$$\bar{u}(z)\frac{\partial\bar{\chi}}{\partial x}=\frac{\partial}{\partial z}\left\{K(z)\frac{\partial\bar{\chi}}{\partial z}\right\} \quad . \quad . \quad . \quad . \quad 5{\cdot}31$$

which with $\bar{u}(z)=\bar{u}_1(z/z_1)^{n/(2-n)}$

and $K(z)=a(n)\bar{u}_1{}^{1-n}z^{2(1-n)/(2-n)}\nu^n, \ z_1=1$

becomes

$$\frac{\bar{u}_1{}^n}{a(n)\nu^n}\frac{\partial\bar{\chi}}{\partial x}=z^{-m}\frac{\partial}{\partial z}\left(z^{1-m}\frac{\partial\bar{\chi}}{\partial z}\right) \quad . \quad . \quad . \quad 5{\cdot}32$$

where $m=n/(2-n)$. This is the standard two-dimensional form of the equation of diffusion for the steady state $(\partial\bar{\chi}/\partial t=0)$.

For a steady continuous infinite line source of strength Q situated along $x=z=0$ the boundary conditions are:

(i) $\bar{\chi}(x, z)\rightarrow 0$ as $x\rightarrow\infty$, $z\geqslant 0$

(ii) $K(z)\partial\bar{\chi}/\partial z\rightarrow 0$ as $z\rightarrow 0$, $x>0$ (i.e. the ground is impervious to the gas)

(iii) $\bar{\chi}(x, z)\rightarrow\infty$ at the line of emission $x=z=0$.

A further condition is imposed by the equation of continuity, i.e. the rate of transfer of matter across unit width of all planes, x=constant is equal to the strength of the source Q, which implies

(iv) $\displaystyle\int_0^\infty \bar{u}(z)\bar{\chi}(x, z)\, dz=Q, \qquad x>0$

Equation 5.32 has been discussed in a rigorous manner by W. G. L. Sutton (1943), with particular reference to the types of boundary conditions which arise in the problem of evaporation (see below). The solution for the present problem was first obtained by O. F. T. Roberts (unpublished) and is (taking $z_1=1$ for convenience and writing $a(n)\nu^n=\alpha$),

$$\bar{\chi}(x,z)=\frac{Q\exp\left\{-\dfrac{\bar{u}_1{}^n z^{(2+n)/(2-n)}}{\left(\dfrac{2+n}{2-n}\right)^2\alpha x}\right\}}{\left(\dfrac{2+n}{2-n}\right)^{(2-n)/(2+n)}\Gamma\left(\dfrac{2}{2+n}\right)\alpha^{2/(2+n)}\bar{u}_1{}^{(2-n)/(2+n)}x^{2/(2+n)}} \quad 5{\cdot}33$$

This expression may be compared with observation from two aspects: first, as regards its functional form, and, secondly, with reference to the absolute magnitudes involved.

In Chapter II it was shown that Scrase found the velocity profile between 3 m. and 13 m. over downland in adiabatic gradient conditions to be $\bar{u}=\bar{u}\,(z/z_1)^{0{\cdot}13}$. This implies $n=0{\cdot}23$. Substituting this value of n in the formula 5·33 it is found that the ground concentration should decay as $x^{-0{\cdot}9}$ which agrees exactly with observation. Other tests (e.g. on the shape of the concentration-height curve) are also satisfactory, so that expression 5·33 is of the correct functional form.

Comparison of the absolute magnitudes, however, reveals a very large discrepancy. This is clearly shown by a calculation of the height of the cloud (Z) at distance x downwind. From 5·33 this is

$$Z=\left\{\log_e 10\left(\frac{2+n}{2-n}\right)^2\alpha\right\}^{(2-n)/(2+n)}x^{(2-n)/(2+n)}\bar{u}_1{}^{-n(2-n)/(2+n)} \quad 5{\cdot}34$$

which for $n=0{\cdot}23$, $k=0{\cdot}4$ becomes

$$Z=0{\cdot}6x^{0{\cdot}79}(\nu/\bar{u}_1)^{0{\cdot}18}\text{ approx.} \quad . \quad . \quad 5{\cdot}341$$

This expression is practically identical with the empirical formula used in aerodynamics for the thickness of the turbulent boundary layer over a smooth rigid boundary, viz.

$$\delta=0{\cdot}37x^{0{\cdot}8}(\nu/\bar{u}_1)^{0{\cdot}2}$$

but for $x=100$ m., $\bar{u}_1=5$ m.sec.$^{-1}$ equation 5·341 gives

$z = 1{\cdot}7$ m. approximately, which is in marked disagreement with the observed height of 10 m. at this distance.

It is now reasonably certain that the failure of the theoretical work to produce results of the correct absolute magnitude can be attributed to the fact that the surface of the earth is almost invariably aerodynamically rough, and quite certainly so in the case of the open downland used for the diffusion experiments. This point is considered in detail later.

Extension of the Statistical Theory to Three Dimensions. The above theory is restricted to diffusion in the vertical and is thus only applicable to sources which are extended infinitely across wind. The statistical theory was adopted for the three-dimensional case by Sutton (1932), who considered the simpler problem of diffusion in a wind constant with height but with the parameter n determined, as before, from observations of the velocity profile. The inconsistency thus introduced has a negligible effect on the absolute value of the concentrations (Sutton, 1947) and allows a very simple treatment of the problem to be developed.

The method consists in finding functions which satisfy Taylor's equation 4.34, the boundary conditions and the equation of continuity. To allow for the non-isotropic character of turbulence near the ground three similar but distinct correlation coefficients are introduced, corresponding to motion along the three principal axes. These coefficients are defined as follows:

$$R_\xi(x)=\left(\frac{\nu}{\nu+\overline{u'^2}\xi}\right)^n;\; R_\xi(y)=\left(\frac{\nu}{\nu+\overline{v'^2}\xi}\right)^n;\; R_\xi(z)=\left(\frac{\nu}{\nu+\overline{w'^2}\xi}\right)^n \qquad 5{\cdot}35$$

The expressions obtained in this way for the continuous point and line sources are:

Continuous point source

$$\bar{\chi}(x, y, z)=\frac{Q}{\pi C_y C_z \bar{u}_1 x^{2-n}} \exp\left\{-\frac{1}{x^{2-n}}\left(\frac{y^2}{C_y^2}+\frac{z^2}{C_z^2}\right)\right\} \qquad 5{\cdot}36$$

Continuous infinite line source

$$\bar{\chi}(x, z)=\frac{Q}{\pi^{\frac{1}{2}} C_z \bar{u}_1 x^{1-\frac{1}{2}n}} \exp\left(-z^2/C_z^2 x^{2-n}\right) \qquad 5{\cdot}37$$

where C_y and C_z are generalized eddy diffusion coefficients defined by

$$C_y^2 = \frac{4\nu^n}{(1-n)(2-n)\bar{u}_1{}^n} g_y{}^{2(1-n)} \quad . \quad . \quad . \quad . \quad 5{\cdot}38$$

$$C_z^2 = \frac{4\nu^n}{(1-n)(2-n)\bar{u}_1{}^n} g_z^{2(1-n)} \quad . \quad . \quad . \quad . \quad 5{\cdot}39$$

g_y and g_z being the components of gustiness defined in Chapter II, p. 16.

These expressions enable the properties of smoke and gas clouds to be calculated from a knowledge of the mean wind at a fixed height ($\bar{u}_1$), the gustiness components (g_y, g_z) and the parameter n and are thus eminently suitable for practical work in the field, since all these quantities can be measured without great difficulty by transportable instruments. Taking, as before, Scrase's velocity profile for adiabatic gradients, i.e. $n{=}0{\cdot}23$, it follows, from equations 5.36 and 5.37 that the peak concentration should decrease downwind as $x^{-1{\cdot}77}$ for the continuous point source and as $x^{-0{\cdot}89}$ for the infinite line source, results which are in good agreement with observations. The expressions are thus of the correct functional form.

The absolute values of the concentration involve the values of the gustiness. These are most conveniently determined in the field from bi-directional vane records, and if the values obtained by Scrase (op. cit.) are used together with $n{=}0{\cdot}23$, the formulae 5.36 and 5.37 are in close agreement with observation (Sutton, 1947). However (as was pointed out by Scrase), these values really represent the upper bounds of gustiness as determined by the instrument and as such must be considerably in excess of the true root mean squares of the oscillations. When more exact estimates of gustiness are used, the theoretical concentrations are much too large, i.e. the predicted rate of diffusion is much too small. A further difficulty arises in the determination of n from the velocity profiles. Above a short cropped grass surface variations in n are chiefly due to changes in temperature gradient, so that in this case the velocity profile can be used as a unique indicator of the stability of the atmosphere. Formulae 5.36, &c., can then be used to predict diffusion in the field over a wide range of conditions, varying from lapse to inversion, but in practice it was found that variations of the same order of magnitude could also arise from changes in the nature of the surface (i.e. of length of grass, &c.), especially when the observations were restricted to the lowest 2 metres of the atmosphere. Here, again, the failure of the theoretical work to produce results of the correct absolute magnitude can be attributed to the fact that the concepts used are essentially those appropriate to flow over smooth surfaces, while the experimental data were obtained from trials over rough uncultivated downland. Despite these imperfections, the formulae developed proved of considerable use in the analysis of experimental data on eddy diffusion.*

4. EVAPORATION FROM A SMOOTH SATURATED SURFACE

The determination of the rate of evaporation of a volatile liquid (such as water) from a plane smooth surface which is infinite across wind but of finite extent downwind involves the solution of the diffusion equation 5.32 subject to the appropriate boundary conditions.† These conditions, however, are by no means obvious, and it is necessary to consider carefully the mechanism of the processes involved in the removal of vapour from a surface in order to ascertain the correct form to be used.

When a slab of porous material is saturated with liquid and then exposed to a current of air, it is found that the drying curve has the general shape shown in Fig. 4. Initially the rate of evaporation is constant, but as drying proceeds a state is reached when the supply of liquid from

* For practical work in connection with chemical warfare, formulae 5.36 and 5.37 have been used as the basis for the design of an ingenious concentration-range slide rule (due to E. Ll. Davies). For this purpose typical values of the 'extreme gustiness' were used, the state of the atmosphere being specified by the values of n found from observations on the velocity profile.

† An alternative method, leading to the same results but using an integral equation, has been produced by K. L. Calder.

the interior of the slab to the surface is insufficient to maintain the rate of evaporation at its original value. The section *AB* of the drying curve is called the 'constant rate period' and *BC* the 'falling rate period'. In this account we shall be concerned only with the constant rate period. Very little progress has been made with the mathematical theory of the falling rate period and the problem is extremely com-

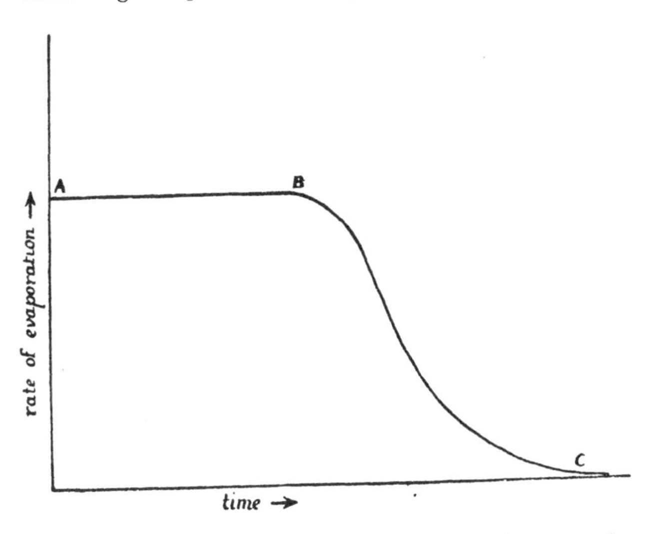

FIG. 4.—Evaporation from a porous slab, initially saturated

plex, involving as it does the mechanism of the removal of vapour by the air stream and of the diffusion of both the liquid and vapour phases in the solid.

In considering the probable form of the conditions which obtain at the boundary during the constant rate period, it is clear from general physical considerations that the mean concentration of vapour $\bar{\chi}(x, z)$ in the air stream will increase steadily as the surface is approached. This increase, however, cannot continue indefinitely because of the limit set by the saturation value. Taking the evaporating surface to be part of the plane $z=0$, in general, three types of surface condition suggest themselves, namely:

(i) the value of $\bar{\chi}$ is prescribed, e.g.

$$\lim_{z\to 0} \bar{\chi}\,(x, z)=\text{const.}=\bar{\chi}_s, \text{ say}$$

(ii) the local rate of evaporation is prescribed, i.e.

$$\lim_{z\to 0} \{K(z)\partial\bar{\chi}/\partial z\}=f(x, \bar{u}_1, \&c.)$$

(iii) the local rate of evaporation and the value of $\bar{\chi}$ at the surface are supposed connected by a linear relation.

A boundary condition of the type (ii) was employed by Giblett (1921), who used as the surface condition an empirical relation (due to Bigelow) between the rate of evaporation, vapour pressure and wind speed. This, however, imposes undesirable and unnecessary restrictions on the generality of the treatment. A condition of the type (iii), which resembles that used in the theory of conduction of heat when radiation is included, does not seem to have been considered by any writer.

In the treatment of the problem due to O. G. Sutton (1934) the surface condition chosen is (i) above, the constant concentration being identified as that corresponding to the saturation vapour pressure. The steady state problem of evaporation then amounts to the determination of the strength of the area source necessary to maintain saturation at the surface itself (which may be solid or liquid) despite the removal of the vapour by the turbulent air stream above.

The problem may now be formulated mathematically.

We require the solution of the equation

$$\frac{\bar{u}_1{}^n}{\alpha z_1{}^m}\frac{\partial\bar{\chi}}{\partial x}=z^{-m}\frac{\partial}{\partial z}\left(z^{1-m}\frac{\partial\bar{\chi}}{\partial z}\right) \quad . \quad . \quad . \quad 5{\cdot}41$$

subject to the boundary conditions

(i) $\lim\limits_{z\to 0} \bar{\chi}\,(x, z)=\bar{\chi}_s=\text{const.} \qquad 0<x\leqslant x_0$

(ii) $\lim\limits_{x\to 0} \bar{\chi}\,(x, z)=0 \qquad 0<z$

(iii) $\lim\limits_{z\to \infty} \bar{\chi}\,(x, z)=0 \qquad 0\leqslant x\leqslant x_0$

where the wetted surface extends from $x=0$ to $x=x_0$ on the plane $z=0$. These suffice to specify the problem for the saturated surface and the region above it. In the case of initially dry air travelling first over the saturated strip and then over dry ground impermeable to vapour (e.g. air blowing from sea to land), the problem is completely specified by conditions (i) and (ii), with condition (iii) replaced by $\lim_{z\to 0}(z^{1-m}\partial\bar{\chi}/\partial z)=0$ for $x>x_0$, i.e. evaporation ceases on the impermeable dry ground.*

The problem constituted by the equation and the boundary conditions can be solved in a number of ways. In his original paper (1934) O. G. Sutton gave the solution in terms of an infinite integral involving Bessel functions; this was later reduced by Whipple to an incomplete gamma function. W. G. L. Sutton (loc. cit.) and R. Frost (1946) also obtained the solution in the incomplete gamma function form, the former using a method based on Goursat's treatment of the classical equation of conduction of heat. Recently J. C. Jaeger (1946) has given a very concise solution of the problem by the method of the Laplace transform.

If $\bar{\chi}(x, z)$ is known the total rate of evaporation can be found by integrating the local rate of evaporation $[K(z)\partial\bar{\chi}/\partial z]_{z=0}$ over the area, but it is also possible to determine the essential facts relating to evaporation very simply and without solving the equation for $\bar{\chi}$. The method is as follows: the transformations

$$\xi=\tfrac{1}{4}(2m+1)^2x/x_0, \qquad \zeta=(\bar{u}_1{}^n/\alpha z_1{}^m x_0)^{1/2}z^{m+1/2}$$

change equation 5.41 into

$$\frac{\partial\bar{\chi}}{\partial\xi}=\frac{\partial^2\bar{\chi}}{\partial\zeta^2}+\frac{1}{(2m+1)\zeta}\frac{\partial\bar{\chi}}{\partial\zeta} \quad . \quad . \quad . \quad . \quad 5.42$$

with the evaporating surface now defined by

$$0\leqslant\xi\leqslant\tfrac{1}{4}(2m+1)^2;\ \zeta=0$$

Thus none of α, $\bar{u}_1$ or x_0 occurs explicitly in the problem

* For a more complete discussion of the mathematical form of the boundary conditions, see W. G. L. Sutton, loc. cit., pp. 52 and 53.

constituted by the transformed equation 5.42 and the transformed boundary conditions, so that if $F(\xi, \zeta)$ is the solution of the latter problem, the solution of the original problem must be

$$\bar{\chi}(x, z)=F\left\{\frac{(2m+1)^2x}{4x_0}, \left(\frac{\bar{u}_1{}^n}{\alpha z_1{}^m x_0}\right)^{1/2} z^{m+1/2}\right\}.$$

The total rate of evaporation, E, is the amount of vapour passing over the plane $x=x_0$, i.e.

$$E=\int_0^\infty \bar{u}(z)\bar{\chi}(x_0, z)\, dz$$

$$=\int_0^\infty \bar{u}_1 z^m F\left\{\frac{(2m+1)^2}{4}, \left(\frac{\bar{u}_1{}^n}{\alpha z_1{}^m x_0}\right)^{1/2} z^{m+1/2}\right\} dz$$

$$=\bar{u}_1\left(\frac{\bar{u}_1{}^n}{\alpha z_1{}^m x_0}\right)^{-(1+m)/(2m+1)}\int_0^\infty F\left\{\frac{(2m+1)^2}{4}, \zeta\right\}\frac{d\zeta}{\zeta^{2(m+1)/(2m+1)}}$$

$$=C\bar{u}_1{}^{(2-n)/(2+n)}\alpha^{2/(2+n)}x_0{}^{2/(2+n)}$$

where C is a function of n and z_1. For $n=\frac{1}{4}$, corresponding to the 'seventh-root profile' we have

$$E=C\bar{u}_1{}^{0\cdot78}\alpha^{0\cdot89}x_0{}^{0\cdot89} \quad . \quad . \quad . \quad . \quad . \quad 5\cdot43$$

This result is in excellent agreement with laboratory data (Sutton, 1934; Lettau, 1937) and brings out the fact that the total rate of evaporation is not exactly proportional to the length of the wetted strip downwind, i.e. the local rate of evaporation must decrease with distance from the windward edge of the strip. (The physical reason for this is, of course, that the air stream is growing richer in vapour as the length downwind of the saturated strip increases.) Evaporation, in fact, is not proportional to the area exposed and the phrase 'rate of evaporation per unit area' is meaningless without qualification. Failure to realize this fact has been responsible for apparent discrepancies in the results obtained by some experimenters.

Equation 5.43 does not permit the calculation of the absolute rate of evaporation *ab initio*; to do this it is necessary to find $\bar{\chi}(x, z)$. The required solutions are as follows:

Vapour concentration over the saturated surface

$$\bar{\chi}(x, z) = \bar{\chi}_s \left[1 - \frac{1}{\pi} \sin \frac{2\pi}{2+n} \Gamma\left(\frac{2}{2+n}\right) \Gamma\left\{ \frac{\bar{u}_1{}^n z^{(2+n)/(2-n)}}{\left(\frac{2+n}{2-n}\right)^2 \alpha z_1{}^{n/(2-n)} x}, \frac{n}{2+n} \right\} \right]$$

where $\Gamma(\theta, p)$ is the incomplete gamma function defined by

$$\Gamma(\theta, p) = \int_0^\theta t^{p-1} e^{-t}\, dt$$

The rates of evaporation are (neglecting lateral diffusion):

Rectangular area $x_0 \times y_0$

$$E(x_0, y_0) = C \bar{u}_1{}^{(2-n)/)(2+n)} x_0{}^{2/(2+n)} y_0$$

where

$$C = \bar{\chi}_s \left(\frac{2+n}{2-n}\right)^{(2-n)/(2+n)} \left(\frac{2+n}{2\pi}\right) \sin \frac{2\pi}{2+n} \Gamma\left(\frac{2}{2+n}\right) \alpha^{2/(2+n)} z_1{}^{-n^2/(4-n^2)}$$

Circular area, radius r_0

$$E(r_0) = C' \bar{u}_1{}^{(2-n)/(2+n)} r_0{}^{(4+n)/(2+n)}$$

where

$$C' = \frac{2^{2+n} \pi^{1/2} \Gamma\left(\frac{3+n}{2+n}\right)}{\Gamma\left(\frac{8+3n}{4+2n}\right)} C$$

The functions C and C' have been tabulated by Pasquill (1943) for $\nu = 0{\cdot}147$ cm.^{2}sec.$^{-1}$, $k = 0{\cdot}4$, $z_1 = 1$ cm. and $0{\cdot}20 \leqslant n \leqslant 0{\cdot}30$.

Comparison with Experiment (*Wind Tunnel*). The above formulae can be evaluated once the value of n has been determined from observations on the velocity profile. A very complete test of the theory has been undertaken by

Pasquill (loc. cit.), who used a variety of liquids (bromobenzene, aniline, methyl salicylate and water) and both rectangular and circular saturated plane surfaces in a specially designed wind tunnel at Porton. The velocity profiles over the surfaces were measured simultaneously and the calculations carried out with the values of n so determined. The results for the evaporation of bromobenzene, expressed in the form E/p_s, where p_s is the saturation vapour pressure and reduced, for convenience of comparison, to a common unit, namely, the value calculated from the theory for $\bar{u}(z_1)=5$ m.sec.$^{-1}$ are as follows:

Comparison of Theoretical and Observed Results for Evaporation (Wind Tunnel)

$u(z_1)$	1	2	3	4	5	6	7	8	9 m.sec.$^{-1}$
Theory	0·28	0·48	0·67	0·84	1·00	1·16	1·31	1·45	1·59
Experiment	0·25	0·43	0·60	0·76	0·92	1·06	1·20	1·34	1·48

Thus the theory slightly overestimates the rate of evaporation but, on the whole, the agreement is as good as can be expected in view of the inevitable limitations of the theoretical approach and may be regarded as a substantial verification of the mathematical method.

Evaporation from Small Saturated Surface in the Open. The wind-tunnel work has been extended to the open air, using again small saturated areas, mounted on a streamlined wooden surround in the middle of a closely cut field. The velocity profile was determined over the adjacent natural surface by the use of special sensitive anemometers at heights from 24 cm. to 200 cm. and the rate of evaporation was determined by weighing. The results are somewhat surprising, for it was found that the rate of evaporation from a small saturated surface at ground level is virtually independent of the stability of the atmosphere as indicated by the temperature gradient, the observed rates of evaporation being, in fact, invariably in close agreement with the theoretical formulae for $k=0{\cdot}4$, $n=\frac{1}{4}$ approx., i.e. the accepted adiabatic gradient values (Sutton, 1947). This result is, of course, quite different from that found in large-scale diffusion (e.g. with smoke clouds) where the rate of diffusion is markedly affected by the stability of the atmosphere.

The only explanation which has been offered so far is that evaporation from a small plate (*c.* 20 cm. diameter) in the open must reflect primarily the nature of the boundary layer set up by the test surface and not the characteristics of the earth's turbulent boundary layer. (With the small plate, the diffusion of vapour occurs in a layer whose depth over the plate is at the most a matter of millimetres.) There is also evidence that, in very close proximity to the ground, flow conditions tend to be largely independent of the thermal stratification of the atmosphere, with the velocity profile always approximating to the adiabatic gradient type. When, however, the mass boundary layer from an evaporating surface is many metres in depth (e.g. evaporation from a lake or large reservoir), it is probable that the rate of loss of vapour from the surface is influenced by the stability of the ambient atmosphere.

These observations have repercussions on the methods adopted by meteorologists for routine measurements of evaporation and it thus appears that the conventional 'evaporimeter' (usually a small tank or saturated surface) is not necessarily a reliable guide to the rate of evaporation from large areas. The whole matter, however, must be regarded as *sub judice* at present and will doubtless remain so until accurate observations on the rate of evaporation from large areas are available.*

5. DIFFUSION IN AERODYNAMICALLY ROUGH FLOW

The work outlined above is essentially dependent upon the hypothesis of the similarity of the diffusion of matter and of momentum, the method adopted being to use the observed velocity profile to predict the rate of diffusion of

* See *Q.J. Roy. Meteor. Soc.*, **73** (1947), pp. 276–81.

matter. So far the analysis has been confined to aerodynamically smooth flow, and it is now necessary to consider what modifications to the theory are necessary to take account of the fact that air flow near the ground is almost invariably of the 'rough' type.

The analogy between eddy viscosity and eddy diffusivity can be considered in greater detail as follows (Calder, unpublished). Taking the logarithmic profile for fully rough flow over a surface with large irregularities

$$\bar{u}/v_* = (1/k) \log_e \{(z-d)/z_0\} \quad . \quad . \quad . \quad 5.51$$

we have the following representative values of the parameters in adiabatic gradient conditions:

Long grass (60–70 cm.): $v_*=50$ cm./sec., $z_0=3$ cm., $d=30$ cm.
Short grass (1–3 cm.): $v_*=33$ cm./sec., $z_0=0\cdot5$ cm., $d=0$ cm.
} for $\bar{u}=5$ m. sec.$^{-1}$ at $z=2$ m.

The use of a logarithmic profile in the equation of diffusion leads to considerable mathematical difficulties which have not yet been overcome and the logarithmic function is accordingly replaced by the power law

$$\bar{u}/v_* = q\{(z-d)/z_0\}^p \quad . \quad . \quad . \quad . \quad 5.52$$

where p and q have to be determined to give a close agreement between equations 5.51 and 5.52. Strictly, this condition implies that both p and q must be variable with height, because a power cannot be made to agree with a logarithm for all values of the variable, but in the subsequent mathematical work it is possible to treat p and q as constants without serious error. When the equivalent power-law wind profile has been determined $K(z)$ can be found explicitly as a power-law function of z, $\bar{u}$, p, q, z_0 and d by the use of Schmidt's conjugate power-law theorem (Chapter III, p. 42). This in turn enables the two-dimensional equation of diffusion to be solved for the case of an infinite line source or for the problem of evaporation, exactly as before.

For the determination of absolute values it is necessary to make a preliminary estimate of the depth of the layer concerned (e.g. of the height of the cloud in the case of an infinite line source of smoke) in order to ascertain the appropriate mean values of p and q. This need only be done very roughly, because both p and q vary quite slowly with the depth of the layer provided that this exceeds about 5 m.

The close agreement between theory and practice which can be attained by this method is shown in the following table (Calder, op. cit.):

Diffusion of Smoke in Adiabatic Conditions calculated from the Diffusion of Momentum (Long Grass Surface)

Source strength, 1 g.m.$^{-1}$sec.$^{-1}$. Wind velocity at 2 m.=5 m.sec.$^{-1}$.

Property of cloud	Calculated	Observed
Height of cloud at 100 m. downwind of source.	10·5 m.	10 m.
Concentration at ground level at 100 m. downwind of source .	35·4 mgm.$^{-3}$	35·6 mg.m.$^{-3}$
Variation of concentration with distance (100–1,000 m.) . .	$x^{-0.86}$	$x^{-0.9}$
Law of variation of concentration in vertical	exp $(-az^{1.37})$	exp $(-az^{p})$ $1.2 \leqslant p \leqslant 1.5$

The agreement is practically perfect and shows that in adiabatic gradient conditions there is complete similarity between the diffusion of momentum and of mass. Essentially, this method is a return to the 'bulk property' approach of Chapter III, in that no use is made of concepts such as mixing length or of correlation between eddy velocities, but consists entirely in the application of the diffusion of momentum to that of mass. The method is therefore subject to the inevitable limitations of this type of analysis in that extensions to three dimensions or to gradients other than the adiabatic are not immediately obvious, but the investigation shows very clearly that in conditions of small vertical temperature gradient (at least) the diffusion of mass can be very accurately predicted from a close study of the velocity profile near the ground. It may also be noted here that if, in the expressions for R_ξ (eq. 5.21) we replace the kinematic viscosity ν by the macro-viscosity $N=v_*z_0$ which occurs in the generalized logarithmic profile for fully rough flow, a fairly close approximation to the absolute value of diffusion over normal terrain is also obtained. The appearance of N in the formula for R_ξ can be justified by arguments similar to those used in deducing the original smooth surface form.

The Stack Problem of Atmospheric Pollution. The theory of diffusion given above has also been applied, with fair success, to the problem of the diffusion of smoke from an elevated source (such as a factory chimney, which may be a hundred metres or more above ground). Diffusion of the plume of smoke from a high stack begins a wind of relatively small gustiness, but the smoke, as it approaches the ground, is subject to two effects: (i) the increasing turbulence of the wind in the lower layers and (ii) the influence of the impervious boundary, acting as a 'reflecting' surface. It is difficult to take full account of the first of these factors and the only practicable method is to assign a mean value for the diffusion coefficient over the height of the stack. The effect of the ground is taken into account by employing the 'method of images', as in corresponding problems in the conduction of heat.

In the treatment of the problem given by Sutton (1947) the following results are obtained for a continuous point source of strength Q at height h above the ground

$$\text{Maximum concentration at ground level} = \frac{2Q}{e\pi u h^2}\left(\frac{C_z}{C_y}\right)$$

$$\text{Distance of point of maximum concentration on ground from foot of stack} = \left(\frac{h^2}{C_z^2}\right)^{1/(2-n)}$$

where n, C_y and C_z have the meanings assigned in equations 5.25, 5.38 and 5.39. The concentration of smoke on the ground is zero at the foot of the stack, rises to a maximum (which varies inversely as the square of the height of the source) and afterwards declines to zero with increasing distance downwind. The result shows that even a relatively small increase in stack height can bring about a marked decrease in the concentration of smoke at ground level. This fact is of importance in studies of atmospheric pollution.

The above expressions have been examined by Gosline (1952) who concludes that they are in sufficiently good agreement with observations to justify their use in practical problems of atmospheric pollution. The main difficulty lies in estimating the increase of concentration to be expected in conditions of marked stability. Sutton has shown that if an inversion exists, the point of maximum concentration will move downwind and that much higher concentrations will be found, but it is not known how far these results are to be trusted quantitatively.

6. DIFFUSION IN THE SIMILARITY THEORY OF TURBULENCE

The methods described in the previous sections are successful in that they establish a technique whereby certain practical problems of meteorology can be solved with tolerable accuracy. On the other hand, such methods throw comparatively little light on the deeper aspects of the problem. For this we must go to theories which have been evolved in recent years, and in particular to the similarity theory of Kolmogoroff (see Chapter IV) and its development by Batchelor.

In 1926 L. F. Richardson pointed out that the characteristic feature of Fickian diffusion is that the rate of separation of two marked particles is, on the average, independent of the distance (l) between them, and he suggested that eddy diffusion differs from molecular diffusion chiefly in that the rate of separation of two particles *increases regularly* with their distance apart. Richardson's analysis involved a kind of diffusion coefficient which he found to be proportional to the $\frac{4}{3}$-power of l. This treatment, called by Richardson the 'distance-neighbour theory', did not attract the attention it deserved at the time, no doubt because of the difficulty of applying Richardson's somewhat abstract reasoning to practical problems. In 1948 Richardson and H. Stommel

7

reopened the subject by studying the diffusion of floating objects on the surface of a Scottish loch and they claimed that the $\frac{4}{3}$-power law was in accordance with the theories of turbulence advanced by Heisenberg and Weiszäcker.

The most satisfactory discussion of the problem on these lines is that of Batchelor (1950), on the basis of the Kolmogoroff theory. Batchelor shows that the problem of diffusion from a fixed point source in a wind is not amenable, in general, to the methods of the Kolmogoroff theory, but that these concepts can be applied to the growth of a cloud of particles drifting with the wind (i.e. to a 'puff' of smoke), provided that the interval of diffusion is not long enough to allow the behaviour of the particles to be dominated by the largest eddies. Batchelor claims that there are good grounds for believing that much of Richardson's early work on diffusion (the work which, in fact, first demonstrated that K increases with the scale of the phenomenon—see p.36) is consistent with the predictions of Kolmogoroff's theory. In particular, it appears that there are good grounds for believing that the virtual coefficient of diffusion increases regularly with the distance apart of the particles, as Richardson indicated.

There is considerable need for basic work of this kind, which makes no use of the somewhat dubious concept of mixing length, but there is the considerable difficulty in that it has proved almost impossible to obtain direct evidence on the rate of separation of floating bodies in the atmosphere. Attempts have been made to do this by following balloons by radar, but so far the results have been inconclusive.

REFERENCES

Sutton, O. G. 1932. *Proc. Roy. Soc.*, A, 135
1934. *Proc. Roy. Soc.*, A, 146
1947. *Q.J. Roy. Meteor. Soc.*, 73
Sutton, W. G. L. 1943. *Proc. Roy. Soc.*, A, 182
Frost, R. 1946. *Proc. Roy. Soc.*, A, 186
Jaeger, J. C. 1946. *Quarterly of App. Maths.*, III
Pasquill, F. 1943. *Proc. Roy. Soc.*, A, 182
Lettau, H. 1937. *Ann. Hydr.*, 65
Gosline, C. A. 1952. *Chem. Eng. Prog.*, 48
Richardson, L. F. 1926. *Proc. Roy. Soc.*, A, 110
—and Stommel, H. 1948. *J. Meteorol.*, 5
Batchelor, G. K. 1950. *Q.J. Roy. Meteor. Soc.*, 133

CHAPTER VI

TURBULENCE IN A VARIABLE DENSITY GRADIENT

THE previous chapters have indicated that a semi-empirical but workable theory of turbulence near the ground can be built up for occasions of adiabatic temperature gradients, i.e. for days and nights of overcast skies and moderate or strong winds. It has already been explained that a characteristic feature of the lower atmosphere is the great variability of the vertical gradient of temperature in clear weather, a fact which makes the study of atmospheric turbulence in such conditions a matter of considerable difficulty, both theoretical and experimental. As yet, no mathematical approach to the problems of thermally stratified flow has been shown to be comparable in accuracy with those developed for an atmosphere in neutral vertical equilibrium, and the reason for this is not far to seek. Apart from the intrinsic mathematical difficulties of the problem there is, in meteorological literature, a dearth of accurate simultaneous measurements of velocity and temperature profiles and an almost complete absence of reliable data on diffusion in gradients other than adiabatic. For conditions of neutral equilibrium the investigator, as we have seen, can rely upon the large body of wind tunnel work to guide him in analysing flow near the ground but, apart from one isolated example,* practical wind tunnels in which the vertical gradient of density can be varied do not seem to exist. Data on flow in these conditions come almost entirely from natural sources.

* Namely, the so-called 'hot-cold' tunnel at Göttingen, used by Reichardt (q.v.).

1. THE RICHARDSON NUMBER AND THE CRITERION OF TURBULENCE

The basic investigation on the effects of gravity on turbulence is that of L. F. Richardson (1920, 1925), who enunciated a criterion to decide whether turbulence will subside or increase in a fluid of variable density gradient. In this respect Richardson's work may be compared to that of Reynolds (Chapter I), and it is now becoming increasingly evident that the non-dimensional quantity which occurs in this criterion, the so-called *Richardson number* (*Ri*), is a factor of major importance in meteorology.

Richardson's principle is usually expressed thus: the kinetic energy of the turbulence will increase or decrease, depending upon the balance between the supply of energy made available by the Reynolds stresses and the rate at which work has to be done against gravity by the turbulence. In this form the principle neglects certain possible energy transformations, a fact appreciated by Richardson, who referred to the criterion as one of 'just-no-turbulence' and limited its application to conditions in which the degree of turbulence is small and the flow borders on the laminar state.

The criterion in its original form may be derived thus. Suppose the fluid is one in which $\bar{p}$, ρ and $\bar{T}$ are all functions of height (z) and consider a volume of fluid moving vertically as a result of the turbulence from a level $z-l$ to a new level z, where l is the mixing length. If it rises without mixing the eddy, which originally had the temperature $\bar{T}(z)-l\,\partial\bar{T}/\partial z$, it will change its temperature adiabatically and will reach the level z with the temperature

$$\bar{T}(z)-l(\partial\bar{T}/\partial z+\Gamma)$$

where Γ is the adiabatic lapse rate. At the new level its excess of density over that of its environment is easily seen to be $(1/\bar{T})l\rho(\partial\bar{T}/\partial z+\Gamma)$, so that there is a buoyancy force brought into being, acting downwards, of magnitude $(g\rho l/\bar{T})(\partial\bar{T}/\partial z+\Gamma)$. If w' be the eddy velocity the mean rate at which work has been done against gravity in lifting the mass is therefore $(g\rho\overline{w'l}/\bar{T})(\partial\bar{T}/\partial z+\Gamma)$, and since, by definition, $\overline{w'l}=K_H$, the eddy conductivity, we have

$$K_H\frac{g\rho}{\bar{T}}\left(\frac{\partial\bar{T}}{\partial z}+\Gamma\right) \quad . \quad . \quad . \quad . \quad . \quad 6.11$$

as the expression for the work done per unit volume, which must be at the expense of the energy of the turbulent motion.

On the other hand, the work extracted from the mean motion by the Reynolds stress and which serves to sustain the turbulence is clearly

$$\tau_{zx}\partial\bar{u}/\partial z=K_M\rho(\partial\bar{u}/\partial z)^2 \quad . \quad . \quad . \quad 6.12$$

per unit volume, where K_M is the eddy viscosity. If now $\bar{E}\rho$ denotes the mean turbulent energy per unit volume of the fluid, Richardson's principle states that the time rate of increase of $\bar{E}$ is given by the difference between 6.12 and 6.11 or

$$\frac{\partial\bar{E}}{\partial t}=K_M\left(\frac{\partial\bar{u}}{\partial z}\right)^2-K_H\frac{g}{\bar{T}}\left(\frac{\partial\bar{T}}{\partial z}+\Gamma\right) \quad . \quad . \quad . \quad 6.13$$

$$=K_H\left(\frac{\partial\bar{u}}{\partial z}\right)^2\left\{\frac{K_M}{K_H}-\frac{g(\partial\bar{T}/\partial z+\Gamma)}{T(\partial\bar{u}/\partial z)^2}\right\} \quad . \quad . \quad 6.14$$

Since $K_H(\partial\bar{u}/\partial z)^2$ is essentially positive and different from zero (except in the trivial case $\bar{u}$=constant), the sign of $\partial\bar{E}/\partial t$ depends on whether the Richardson number

$$Ri=\frac{g(\partial\bar{T}/\partial z+\Gamma)}{T(\partial\bar{u}/\partial z)^2} \quad . \quad . \quad . \quad . \quad . \quad 6.15$$

is less or greater than the ratio K_M/K_H. In his original discussion Richardson assumed $K_M=K_H$, in which case the criterion becomes

turbulence increases if $Ri<1$
turbulence decreases if $Ri>1$

implying the existence of a critical value (Ri_{crit}) of the Richardson number, which in this case is unity.

Taylor (1931), using the method of small oscillations, has investigated the similar problem of the stability of a non-viscous fluid of infinite extent in which the velocity gradient (α) is constant and the density gradient (β) is small and uniform. The result obtained is that the system is stable if

$$g\beta/\alpha^2 > 0{\cdot}25$$

the quantity on the left-hand side being equivalent to the Richardson number. (These calculations have also been applied to the case of three and four superposed fluids moving with uniform shear.) Later, Schlichting (1935), by extending the calculations of Tollmien on the stability of a boundary layer in a viscous incompressible fluid, considered the problem for flow near a boundary. He obtained the result that Ri_{crit} depends on both the Reynolds number and the Froude number*; for large Reynolds number and small Froude number—the meteorologically significant case—Schlichting gave the critical value of Ri as 0·04, a result which has been substantially confirmed in the laboratory. Reichardt (1934), working with the Gottingen 'hot-cold' tunnel (in which a stabilizing density gradient in the boundary layer is produced by heating the roof of the tunnel by steam and cooling the lower surface by running water), found good agreement with Schlichting's calculations.

Determinations of the Critical Value of Ri *in the Atmosphere.* The theoretical work has thus indicated a number of possible values of Ri_{crit}, depending upon the type of situation analysed. Unfortunately, the evidence from meteorological investigations is conflicting, but this is not altogether surprising in view of the fact that a precise determination of the conditions in which turbulence dies away requires highly accurate and detailed observations of both the velocity and temperature fields.

Durst (1933) showed that in the lower layers of the atmosphere the magnitudes of the nocturnal inversion and of the wind shear usually increase simultaneously until a limiting value is reached, at which point vigorous oscillations appear

* The Froude number is $\bar{u}/\sqrt{(\delta g)}$, where $\bar{u}$ is the free stream velocity and δ is the depth of the boundary layer.

in the velocity followed by a quick collapse of both the temperature and velocity gradients as a result of the mixing. From the data he concluded that there is good evidence to support Richardson's original value, namely $Ri_{crit}=1$. Paeschke (1937), on the other hand, found that observations on wind and temperature profiles near the surface confirm Schlichting's value of $Ri_{crit}=\frac{1}{25}$. Petterssen and Swinbank (1947), basing their conclusions on observations made in the free atmosphere, suggest that the value of Ri_{crit} is about 0·65. Sutton (unpublished) has shown that during the inversion period the width of the cloud from a continuous point source of smoke, measured at a fixed distance downwind, satisfies the empirical law

$$\text{width}=a-b\sqrt{Ri}$$

where a and b are positive constants, and by extrapolating this equation to zero width (corresponding to a complete absence of eddy diffusivity) finds that $Ri_{crit}=\frac{1}{2}$ approximately. Most recently, Deacon (unpublished), using a pressure tube anemometer and very accurately measured profiles of temperature and velocity, has produced what appears to be exceptionally good evidence that in the layer 0·5–4 m. turbulence becomes vanishingly small when Ri exceeds about 0·15. Thus at present some support can be found from meteorological data for almost any value of Ri_{crit} between 0·04 and 1, an unsatisfactory state of affairs which emphasizes the need for further research work on this fundamental aspect of atmospheric turbulence.

The derivation of the criterion given above is not altogether satisfactory, and a more critical examination of equation 6.13 throws doubt on its general applicability. It has been pointed out in Chapter II that in overcast windy weather the temperature gradient in the lower layers of the atmosphere is invariably very close to the adiabatic lapse rate, so that $\partial\bar{T}/\partial z+\Gamma$ is effectively zero in these conditions. At the same time $\partial\bar{u}/\partial z$ is certainly not zero and may even assume quite large values, especially over rough surfaces, so that, according to equation 6.13, $\partial\bar{E}/\partial t$ is positive, i.e. the kinetic energy of the turbulence should be increasing.

Observations, however, show that in overcast weather the level of turbulence is remarkably steady if the mean wind is steady, suggesting that certain terms which have been disregarded in the derivation of equation 6.13 may become of importance in these circumstances.

The derivation of the Richardson criterion has been re-examined by K. L. Calder (unpublished), using a method based on transformations of the equation of motion, similar to that employed by Reynolds in his classical discussion of the incidence of turbulence in an incompressible fluid. The analysis is too long and intricate to be given here, but the final result may be stated quite simply. Calder finds that in its original form the criterion omits certain terms, the most important of which is that representing the mean rate of working of the eddy velocities against the local fluctuating pressure gradients which necessarily accompany turbulent flow. The validity of Richardson's criterion of 'just-no-turbulence' depends essentially upon this term being small when the degree of turbulence is small, and according to this analysis the general form of the criterion for the extinction of turbulence is

$$Ri_{crit} > 1 - \delta, \qquad 0 < \delta < 1$$

assuming $K_M = K_H$ as before.

The Richardson Number in the Expression of the Velocity Profile. Rossby and Montgomery (1935), by comparing a homogeneous and a stratified medium moving under the same shearing stress have deduced the equation

$$d\bar{u}/dz = (v_*/kz)\sqrt{(1 + \sigma Ri)} \quad . \quad . \quad . \quad 6.16$$

where σ is a positive constant. (This equation should be compared with equation 4.21, p. 53, for a non-stratified fluid.) The analysis by which this expression is derived is, however, open to some doubt. Sverdrup (1936), making the additional assumption that the profiles of velocity and potential temperature are similar in the vicinity of the surface, has shown that the above equation leads to the following limiting cases:

small stability: $d\bar{u}/dz \propto 1/z$

large stability: $d\bar{u}/dz \propto 1/z^{2/3}$

and has suggested that in general the relation

$$d\bar{u}/dz \propto (z+z_0)^{(1-n)/n}, \qquad 3<n<\infty$$

is satisfied in inversion conditions, a conclusion which leads to expressions which agree very well with observations made over a smooth snow surface. (These results should be compared with those found by Deacon, Chapter II, p. 22, for the wind structure over downland.) Sverdrup also evaluated the Rossby-Montgomery stability parameter σ and found its value to be about 11, but it seems likely that σ as defined above is not a true constant but a quantity which decreases in strongly unstable conditions and increases rapidly in stable conditions when Ri exceeds about 0·1.

Equation 6.16 may be interpreted as the form taken by the Prandtl equation $l d\bar{u}/dz = v_* =$const. (Chapter IV) when $l = kz/\sqrt{(1+\sigma Ri)}$. Holzman (1943) has suggested that the Rossby-Montgomery equation 6.16 should be replaced by

$$d\bar{u}/dz = v_*/\{kz\sqrt{(1-\sigma Ri)}\} \quad . \quad . \quad . \quad 6.17$$

which amounts to taking $l = kz\sqrt{(1-\sigma Ri)}$, so that $l=0$ (which may be taken to correspond to the disappearance of turbulence) when $Ri = 1/\sigma$, thus allowing the critical value to be found. Holzman's form of the equation has been examined by Deacon (op. cit., unpublished), who finds that it is superior to the Rossby-Montgomery equation in that the values of σ thus obtained show no systematic variation with roughness or stability but are grouped around a mean value $\sigma=7$, indicating that $Ri_{crit}=0{\cdot}15$ approximately. Equations 6.16 and 6.17 are not very different for small Ri (say $-0{\cdot}1 < Ri < 0{\cdot}1$), but Holzman's formula for l has the additional merit of indicating a definite change of regime at a critical positive value of Ri, at which point, of course, equation 6.17 breaks down.

The existence of a universal relation between $d\bar{u}/dz$ and Ri has been shown convincingly by Deacon (1947) by plotting the ratio $\bar{u}_{4m.}/\bar{u}_{\frac{1}{2}m.}$ against Ri for a short grass surface. The points all fall on substantially the same curve for $-0{\cdot}1 \leqslant Ri \leqslant 0{\cdot}1$.

2. CONVECTION IN THE LOWER ATMOSPHERE

The transfer of heat in the vertical is due to the combined effects of molecular conduction, radiation and convection, the last term being taken to mean the phenomenon of the spread of heat by the movement of relatively large masses of air. Of these, conduction may usually be neglected in meteorological problems, except possibly in the immediate vicinity of the ground. The transfer of heat by radiation is a complex problem, because the atmosphere is crossed by beams of both short-wave (*c.* $0{\cdot}5\,\mu$) and long-wave (*c.* $10\,\mu$) radiation and is selective in its powers of absorption of the latter mainly because of the presence of water-vapour. It is usually assumed that radiation may be disregarded in problems of the transfer of heat near the ground, but the question is one which, as yet, is not satisfactorily settled.

In the mid-hours of a clear summer day it appears to be legitimate to assume that the factor which is chiefly responsible for the transfer of heat in the vertical is convection, either forced or natural, and we shall conclude this chapter with an account of this problem, one of the most intricate in meteorology.

Forced Convection. The experimental study of forced convection in the lower atmosphere, i.e. of the rate of removal of heat from the ground by turbulence in moderate or high winds, is a matter of some difficulty chiefly because in these conditions the mixing is so thorough that the temperature field near the ground tends to become uniform with height and the measurement of the flux of heat from one level to another requires extremely accurate observations. What evidence is available on these problems comes almost entirely from laboratory work, where the actual loss of heat from a surface may be determined without great difficulty.

The question of the similarity or otherwise of the transfer of heat and momentum has been discussed extensively, the most notable contributions being those of Reynolds, Taylor, Prandtl and Kármán. Reynolds assumed the mechanism of the two processes to be identical, resulting in the so-called 'Reynolds' analogy' which for an incompressible fluid may be expressed mathematically by the equations

$$\tau/\rho=-(\nu+K)d\bar{u}/dz$$
$$q/c\rho=-(k'+K)d\bar{T}/dz$$

where q is the flux of heat across the level z, c is the specific heat and k' is the thermometric conductivity ($=\kappa/\rho$ where κ is the thermal conductivity), the other symbols having their usual meaning. The quantity K is the eddy viscosity or conductivity, assumed identical for momentum and heat. Prandtl (1928) showed that the analogy can only be complete if certain conditions are satisfied, these being (i) similarity of boundary conditions, (ii) negligible pressure drop in the direction of flow, and (iii) the non-dimensional quantity $\mu c/\kappa$ (usually termed the 'Prandtl number') must be equal to unity. The problem of the transfer of heat in the presence of a pressure gradient (e.g. in pipes and wakes generally) has been examined by Taylor (1932); in these circumstances the equation of momentum transfer contains a term which has no counterpart in the equation of heat transfer and complete correspondence cannot be expected. This complication does not arise in the atmospheric problem where the change of pressure down wind may usually be neglected.

The loss of heat by forced convection from a smooth flat plate in a wind tunnel has been measured by Éliás (1929, 1930), whose results have been compared with theory by Pasquill (1943). Éliás used a fairly high speed air current parallel to the plate (*c.* 35 m.sec^{-1}) and determined the loss of heat from the surface by measuring the input of electrical energy necessary to maintain the plate at a constant temperature. In these circumstances the velocity and temperature profiles over the plate were found to be practically identical.

The boundary conditions in Éliás's experiments correspond exactly to those assumed by Sutton in his treatment of evaporation (Chapter V, p. 79) with the constant concentration of vapour replaced by the constant temperature, and this fact has been used by Pasquill (loc. cit.) to examine if the total loss of heat from the plate could also have been predicted from the observed diffusion of momentum. Replacing the concentration χ by $c_p\rho T$ and determining n from Éliás's velocity profiles, Pasquill found fair agreement between theory and observation for the rate of loss of heat both as regards the functional form and the absolute value. In this example, which corresponds to the case of a high wind blowing over a smooth surface whose temperature does not differ greatly from that of the air stream, it may be concluded that the rates of transfer of heat and momentum are practically identical. It is also clear that in this case buoyancy forces must have been at a low level compared with frictional forces.

Natural Convection in the Atmosphere. The temperature field in the first hundred metres of the atmosphere on a clear hot day has been described in considerable detail by Johnson and Heywood (1938) in their analysis of temperature gradient observations at Leafield, Oxfordshire. In addition to the rapid fall of temperature with height, the records show characteristic oscillations which Johnson and Heywood ascribe to large masses of air breaking away from the ground and ascending to considerable heights, the process being repeated at intervals of the order of a minute. This observation suggests a method by which an analysis of the effects of natural convection (i.e. convection in a calm or a low wind) can be made.

Suppose that a volume of heated air rises through the atmosphere because of its reduced density and let l be the characteristic linear dimension of the volume. The forces acting on the volume are its buoyancy and the resistance experienced in breaking through the cooler air. If $\delta\theta$ be the excess potential temperature of the volume, the force of buoyancy is proportional to $g\rho l^3\delta\theta/\theta$, while the resistance experienced is proportional to $\rho l^2 w^2$, where w is the velocity of ascent. If the ascending mass adjusts its velocity so as to achieve a balance between these two forces we have

$$\rho l^2 w^2 = \text{const. } g\rho l^3\delta\theta/\theta$$

or $$w^2 = \text{const. } gl\delta\theta/\theta$$

The excess temperature $\delta\theta$ may be taken proportional to $-l\partial\bar{\theta}/\partial z$ and since the potential temperature and the absolute temperature (T) are related by

$$\frac{1}{\bar{\theta}}\frac{\partial\theta}{\partial z} = \frac{1}{\bar{T}}\left(\frac{\partial\bar{T}}{\partial z} + \Gamma\right)$$

it follows that the velocity of ascent is determined by the equation

$$w^2 = \text{const. } \frac{gl^2}{\bar{T}}\left(-\frac{\partial\bar{T}}{\partial z} - \Gamma\right) \quad . \quad . \quad . \quad . \quad 6.21$$

The virtual coefficient of conduction is $K = wl$ so that finally

$$K(z) = l^2\sqrt{\left\{-\frac{g}{\bar{T}}\left(\frac{\partial\bar{T}}{\partial z} + \Gamma\right)\right\}} \quad . \quad . \quad . \quad . \quad 6.22$$

where constants of proportionality may be assumed to be absorbed in the quantity l^2.

We may look at the same problem from another angle. The buoyancy of the heated masses which break away from the surface would give rise to very large values of the vertical velocity and the mixing length were it not for the continual degradation of the kinetic energy of the eddies by the breakdown of the original masses into smaller eddies. Thus the intensity of the convection currents may be considered to be determined principally by a balance between the disintegration of the 'bubbles' of hot air and the rate of loss of potential energy. This observation, as Taylor has shown, leads to another and possibly more satisfactory deduction of formula 6.21 above. The mean rate of loss of eddy motion by viscosity in isotropic turbulence has been shown in Chapter IV, p. 64, to be

$$\overline{W} = \text{const. } (\rho w^3/l)$$

The rate at which potential energy is being lost per unit volume is, as before, $g\rho w\delta\theta/\theta = g\rho wl\left(\frac{1}{\bar{\theta}}\frac{\partial\bar{\theta}}{\partial z}\right)$, if it be assumed that the convection currents only preserve their identity over a fraction of the heated volume. Thus the assumed balance implies

$$\rho w^3/l = \text{const. } g\rho w l\left(\frac{1}{\bar{\theta}}\frac{\partial\bar{\theta}}{\partial z}\right)$$

giving equation 6.21 as before.

These relations take a particularly simple form when the flux of heat (q) is invariable with height, a condition which obtains in the lower atmosphere in the mid-hours of a clear summer day (*vide* Chapter II, p. 26). Since by definition

$$q = -Kc_p\rho(\partial\bar{T}/\partial z + \Gamma)$$

it follows from 6.22 that

$$K^3 = -(g/\bar{T})l^4K(\partial\bar{T}/\partial z + \Gamma)$$

whence

$$K = \text{const. } l^{4/3}$$

and also

$$w = \text{const. } l^{1/3}$$

$$\partial\bar{T}/\partial z + \Gamma = \text{const. } l^{-4/3}$$

$$T' = l(\partial\bar{T}/\partial z + \Gamma) = \text{const. } l^{-1/3}$$

where T' is the oscillation of temperature at height z. Thus a knowledge of the variation with height of any of the quantities K, $\partial\bar{T}/\partial z + \Gamma$ or T' and in particular of l, should, if the above theory is correct, enable the variation with height of all the other quantities to be determined. The truth of the above picture of natural convection thus depends largely on whether the above expressions form a set consistent within themselves and with observation, a matter which fortunately can be fairly easily examined.

Comparison between Theory and Observation. The comparison described above has been made by Sutton (1948), using data given by Johnson and Heywood (op. cit.) for clear June days at Leafield. From records giving the mode of variation of T' and $\partial\bar{T}/\partial z$ with height it was found that, in the hours around noon on a clear June day,

$$l = T'/(\partial\bar{T}/\partial z + \Gamma) = \text{const. } z^{1\cdot 35}, \qquad 7\text{m} \leqslant z \leqslant 45\text{m}$$

Thus if the theory be true, it follows that over the same heights and for the same conditions, $\partial \bar{T}/\partial z+\Gamma$ should decrease as $z^{-1\cdot80}$. The Leafield data show that the law of decrease with height is actually

$$\partial \bar{T}/\partial z+\Gamma=\text{const. } z^{-1\cdot75}, \qquad 7\text{m} \leqslant z \leqslant 45\text{m}$$

which indicates a satisfactory verification of the theoretical relations. It also follows that in the same conditions and over the same range of height

$$K=\text{const. } z^{1\cdot75}, \qquad 7\text{m} \leqslant z \leqslant 45\text{m}$$

This result is in harmony with those obtained by earlier workers using the inexact but convenient method of applying the solution of the conduction equation 3.22 for K=constant to a succession of shallow layers and then expressing K as a function of height, namely, that in clear warm weather the eddy conductivity varies approximately as the square of the height (Chapter III).

The Representation of Natural Convection by a Model. There is no precise information on the form of the convection currents which rise from the heated ground in clear weather, but the matter has elsewhere received attention from physicists and there are indications that the atmosphere shows on a larger scale what has been found in the laboratory for very shallow layers. A mass of fluid heated from below does not necessarily break down into a state of complete disorder but tends initially to divide into 'cells' in which there is a steady upward motion at the centre, a diverging motion at the top and a descending motion in the outer regions. The motion in such cells has been investigated mathematically by Rayleigh, Jeffreys, Low and others, who have shown that a fluid can exhibit permanent motions of this type provided that the density gradient does not exceed a certain limiting value.* A quasi-regularity in the bursts of convection from the ground has been detected by Johnson and Heywood (op. cit.), as indicated above, and we

* For a further discussion of this interesting but difficult problem the reader should consult Brunt, *Physical and Dynamical Meteorology*, Chapter XI. The detailed mathematical theory is given in Jeffreys, *Methods of Mathematical Physics*, pp. 413–15.

may obviously regard the whole process as one in which 'bubbles' of hot air periodically break away from the surface and ascend into the cooler air above. The integrated effect of a large number of such bubbles is probably not unlike that produced by a succession of instantaneous plane sources of heat generated at ground level and spreading vertically as they rise under their own buoyancy. This suggests that a typical convection eddy resembles a puff of smoke, of great horizontal area but initially relatively shallow, which 'grows' as it rises and so spreads its excess of heat over a deeper layer of air. The flux of heat due to such a convection eddy is then made up of two parts: (i) the flux caused by the ascent of the heated masses due to their buoyancy, and (ii) the additional flux due to the heated masses spreading upwards and downwards because of the turbulent mixing generated by the heated air breaking through the cooler quiescent atmosphere.

A model of this type has been devised by Sutton (1948) to account for the observed features of the temperature field in the lower atmosphere in the mid-hours of a clear summer day. It is assumed that inside the bubble the distribution of temperature follows the same law as the distribution of matter in a puff of smoke, but that instead of a wind moving the puff away from the source (the plane of the earth) it now rises with the convectional velocity w. The results show fair agreement with observation, so that it appears probable that the broad features of the temperature field near the surface on a hot day in summer are in the main due to the ascent and diffusion of large isolated masses of heated air.

3. THE PROBLEM OF THE CONVECTIVE JET

Measurements of pollution from industrial sources indicate that two very effective ways of reducing the concentration of smoke downward of a stack are: (i) by increasing the height of the stack (see Chapter V), and (ii) by ensuring that the smoke leaves the stack orifice at as high a temperature as possible. In the second of these methods, advantage is taken of natural convection to increase the effective height of the stack.

Detailed examination of the behaviour of a jet of heated air rising in a *calm* atmosphere of uniform potential temperature has been made by Schmidt (1941) and Sutton (1950). Both treatments depend on mixing-length concepts, but differ in detail. In Schmidt's analysis the final results are expressed as infinite series, but Sutton's method gives results in a closed form suitable for computation.

A jet of hot air rising in a calm atmosphere takes a conical shape, with entrainment of cool air at its boundary. The upward velocity is a maximum at the centre of the jet and declines to zero at the edges, so that the mixing process which entrains the cool air is like that in the turbulent boundary layer adjacent to a solid surface (see Chapter IV). In the jet problem the mixing length is supposed to be proportional to the radius of the jet at any level, and Sutton assumes that the turbulent velocity fluctuations are proportional to the rate of decrease of the velocity of ascent (w) and height (z). From these assumptions it follows that, if Q be the strength of the source (cal./sec.), θ the mean excess temperature in the jet at any level z, T, ρ and c_p absolute temperature, density and specific heat, respectively, of the ambient atmosphere, and C the diffusion coefficient defined in Chapter V,

$$\theta = \frac{Q}{2{\cdot}3\pi c_p C^2 w_1 z^{1{\cdot}46}}$$

$$w = w_1 z^{-0{\cdot}29}$$

$$w_1 = \left(\frac{728{\cdot}5Q}{c_p \rho TC}\right)^{\frac{1}{3}}$$

These expressions show that the excess temperature in the convective jet from a continuous point source of heat falls off fairly rapidly with height (approximately as $z^{-3/2}$) in an atmosphere of uniform potential temperature, but that the upward velocity decreases slowly (approximately as $z^{-1/3}$). The expressions given above have been compared with laboratory observations by Railston (1954), who found good agreement between theory and measurement. No tests have been carried out in the open air, and it is difficult to see how reliable observations could ever be made in such circumstances.

The problem of a convective jet in a side wind is much more complicated and it would be unreasonable to expect an exact solution in the present state of knowledge of turbulence accompanying both dynamical and thermal instability. By considering the problem of a particle subjected to a constant horizontal velocity (the side wind) and an upward velocity which decreases with distance along the trajectory as the cube-root of the height, Sutton has shown that it is probable that the height reached by a plume in a wind varies inversely as the cube of the side wind. (This result affords an explanation of the guttering of a candle flame in a relatively feeble current of air.)

It can also be shown that the effect of adding a source of heat Q to a plume from an elevated source of height h is to reduce the maximum concentration at ground level by an amount proportional to Q/u^3h (Sutton, *loc. cit.*), where u is the horizontal wind. This indicates precisely the value of conserving or adding heat to industrial sources of smoke as a method of reducing atmospheric pollution in the vicinity of the plant. High chimneys, designed to conserve as much heat as possible, are essential if those living in industrial areas are to enjoy a relatively clean atmosphere.

REFERENCES

Richardson, L. F. 1920. *Proc. Roy. Soc.*, A, 97
1925. *Phil. Mag.*, 49
Taylor, G. I. 1931. *Rapports et Proces—Verbaux du Conseil Permanent pour l'Exploration de la Mer*, 76
1931. *Proc. Roy. Soc.*, A, 132
Schlichting, H. 1935. *Zeit. f. angew Math. u. Mech.*, 15
Prandtl and Reichardt. 1934. *Deutsche Forschung*, Part 21
Durst, C. S. 1933. *Q.J. Roy. Meteor. Soc.*, 59
Paeschke, W. 1937. *Beit. Phys. fr. Atm.*, 24
Petterssen, S., and Swinbank, W. C. 1947. *Q.J. Roy. Meteor. Soc.*, 73
Rossby, C., and Montgomery, R. B. 1935. *M.I.T. Pap. Phys. Ocean. Met.*, 3 and 4
Sverdrup, H. U. 1936. *Met. Zeit.*, 53
Holzman, B. 1943. *Ann. New York Acad. Sci.*, 44
Éliàs. 1929. *Zeit. f. angew Math. u. Mech.*, 6
Schmidt, W. 1941. *Zeit. f. angew Math. u. Mech.*, 21, 265, 351
Sutton, O. G. 1948. *Q.J. Roy. Meteor. Soc.*
1950. *J. Meteorol.*, 7, 307
Railston, W. R. 1954. *Proc. Phys. Soc.*, B, 67

APPENDIX

NUMERICAL DATA FOR FLOW NEAR THE SURFACE OF THE EARTH

TABLE I

Typical Values of the Roughness Parameter (z_0), *Drag Coefficient of Surface* (C_d) *and Friction Velocity* (v_*)

(*for a mean wind of* 500 *cm.sec.*$^{-1}$ *at a height of about* 2 *m.*)

($\bar{u}/v_* = (1/k) \log_e (z/z_0)$; $v_*^2 = \tau/\rho = \frac{1}{2} C_d \bar{u}_1^2$)

Type of surface	z_0(cm.)	C_d	v_*(cm.sec.$^{-1}$)
Exceptionally smooth (mud flats, ice)	1×10^{-3}	2×10^{-3}	16
Smooth sea	2×10^{-2}	$3{\cdot}8\times10^{-3}$	21
Level desert (India) . .	3×10^{-2}	$4{\cdot}1\times10^{-3}$	23
Lawn (grass *c.* 1 cm.) .	0·1	$5{\cdot}5\times10^{-3}$	27
Lawn (grass *c.* 5 cm.) .	1–2	$1{\cdot}5\times10^{-2}$	43
Long grass (*c.* 60 cm.) .	4–9	$0{\cdot}3\times10^{-2}$	60
Fully grown root crops .	14	$4{\cdot}3\times10^{-2}$	70
Downland (winter) . .	1–2	$1{\cdot}3\times10^{-2}$	40
Downland (summer) . .	2–4	2×10^{-2}	50

(Data from P. A. Sheppard, *Proc. Roy. Soc.*, A, 188 (1947) and E. L. Deacon [unpublished])

TABLE II

Values of the Diffusion Function a(n) (see Chapter V, equation 5.26)

$$a(n)=\frac{(\frac{1}{2}\pi k^2)^{1-n}(2-n)n^{1-n}}{(1-n)(2-2n)^{2(1-n)}}, \qquad k=0{\cdot}4$$

n	$a(n)$	n	$a(n)$	n	$a(n)$	n	$a(n)$
0·10	0·025	0·20	0·086	0·30	0·212	0·40	0·447
0·11	0·029	0·21	0·095	0·31	0·229	0·41	0·479
0·12	0·034	0·22	0·105	0·32	0·248	0·42	0·513
0·13	0·038	0·23	0·115	0·33	0·268	0·43	0·549
0·14	0·044	0·24	0·127	0·34	0·289	0·44	0·587
0·15	0·050	0·25	0·138	0·35	0·312	0·45	0·628
0·16	0·056	0·26	0·151	0·36	0·336	0·46	0·671
0·17	0·063	0·27	0·165	0·37	0·361	0·47	0·716
0·18	0·070	0·28	0·180	0·38	0·388	0·48	0·764
0·19	0·078	0·29	0·195	0·39	0·417	0·49	0·868

INDEX